RUBBER BAND
ENGINEER

ROCKPORT

RUBBER BAND
ENGINEER

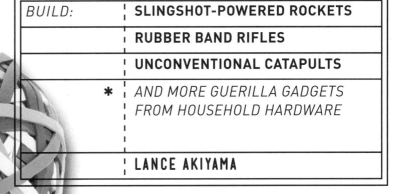

BUILD:		SLINGSHOT-POWERED ROCKETS
		RUBBER BAND RIFLES
		UNCONVENTIONAL CATAPULTS
*		*AND MORE GUERILLA GADGETS FROM HOUSEHOLD HARDWARE*
		LANCE AKIYAMA

ROCKPORT

Quarto is the authority on a wide range of topics.

Quarto educates, entertains and enriches the lives of our readers—enthusiasts and lovers of hands-on living.

www.QuartoKnows.com

First published in the United States of America in
2016 by Rockport Publishers, an imprint of
Quarto Publishing Group USA Inc.
100 Cummings Center
Suite 406-L
Beverly, Massachusetts 01915-6101
Telephone: (978) 282-9590
Fax: (978) 283-2742
QuartoKnows.com
Visit our blogs at QuartoKnows.com

10 9 8 7 6 5 4 3 2 1

ISBN: 978-1-63159-104-4
eISBN: 978-1-63159-180-8

Library of Congress Cataloging-in-Publication Data
available.

Design and illustrations: Timothy Samara
Cover image: Lance Akiyama
All photography by Lance Akiyama except the cover
(rubber band ball) and pages 3 and 8 by Shutterstock

Printed in China

Rubber Band Engineer contains a variety of desktop
projects involving projectiles made from household items.
Neither Quarto Publishing Group USA Inc. nor the author
accepts responsibility for the use or misuse of the information contained in this book.

Thank you to my parents, Nancy and James, for allowing me to take apart old electronics and for letting me turn my childhood bedroom into a wonderful, creative mess. And, thank you to the inventors of rubber bands, cardboard, and tape, without whom I might not have a career.

- -

LANCE

CONTENTS

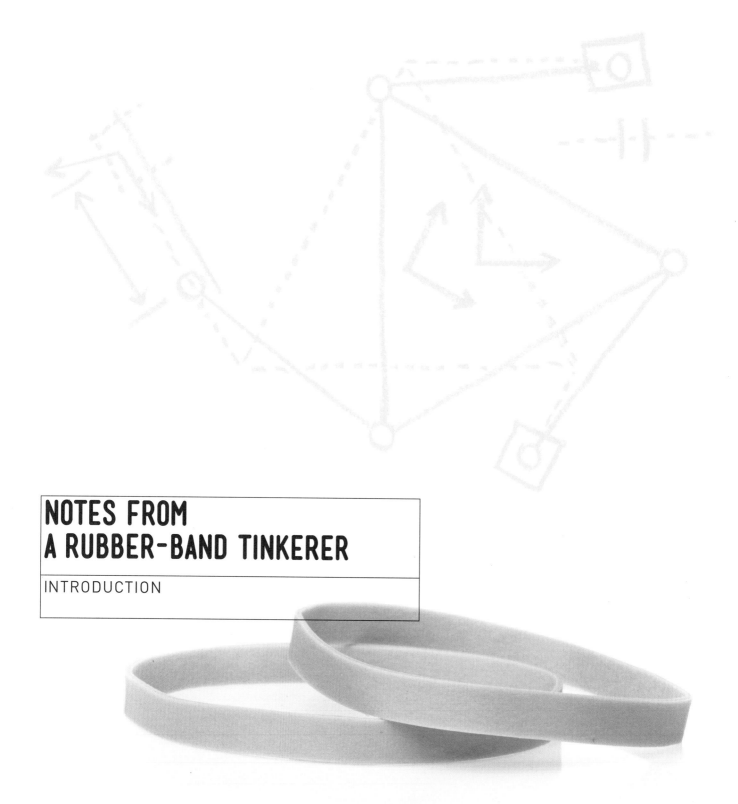

NOTES FROM
A RUBBER-BAND TINKERER

INTRODUCTION

"Whoa, that shot a lot farther than I thought it would!"

Shooting far, flying high, and delivering way more exciting results than expected are the goals for the gadgets in this book.

Discover unexpected ways to turn common materials into crafty contraptions that range from surprisingly simple to curiously complex. Whether you build one or all of these designs, you'll feel like an ingenious engineer when you're through.

Best of all, you don't need to be an experienced tinkerer to make any of these projects. All you need are household tools and materials, such as paper clips, pencils, paint stirrers, and ice pop sticks.

Oh, and rubber bands. *Lots of rubber bands.*

So grab your glue gun, pull out your pliers, track down your tape, and get started on the challenging, fun, and rewarding journey toward becoming a rubber band engineer.

P.S.
RUBBER BAND
ENGINEERS:
IMPORTANT
LEGAL SAFETY
DISCLAIMER

There are things
in this book that
are dangerous.
Be smart. Tinker
at your own risk.

HANDHELD
SHOOTERS

MANY-THING SHOOTER
PVC SLINGSHOT RIFLE
CROSSBOW
BOW AND ARROW

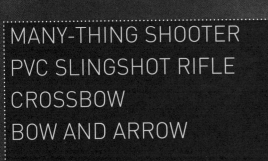

The Many-Thing Shooter earns its name from its versatility: it can be built from many things. It can also launch many things, including candy, beads, and pieces of cork, to name a few. Every material in the shooter can be replaced with another household item, making this one of the most imaginative, adaptable, and quick-to-build projects in this book. Experiment with different projectiles and rubber-band combinations until your Many-Thing Shooter becomes your go-to DIY sidearm.

MANY-THING SHOOTER

$$\Delta x = \bar{v}t$$
$$\Delta x = \frac{1}{2}at + v_0$$
$$v = at + v_0$$

CHOOSE YOUR RUBBER BAND WISELY

The rubber band is a critical component of the shooter. Choose a band that has a maximum stretched length about equal to the length of the stick. Avoid using very thin bands that break easily or very thick bands that are difficult to stretch. Try doubling your rubber bands for additional power!

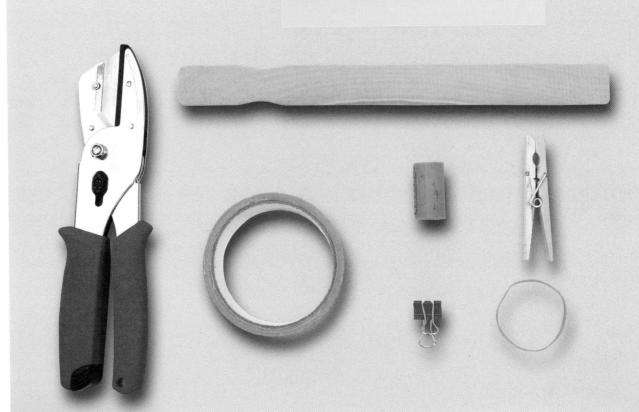

TOOLS + MATERIALS

DUCT TAPE

SCISSORS

PAINT STIRRER

SPRING-TYPE CLOTHESPIN

RUBBER BAND

SMALL BINDER CLIP

**BOTTLE CORK
(OR PROJECTILE OF
YOUR CHOICE)**

CUTTING TOOL (OPTIONAL)

MATERIAL SUBSTITUTIONS

PAINT STIRRER	Ruler, wood shim, craft sticks; or another flat, rigid strip of wood, plastic, or cardboard
CLOTHES PIN	Binder clip, chip clip, or anything that can clamp onto a rubber band and act as a trigger
BINDER CLIP	Masking tape or duct tape
DUCT TAPE	Hot glue, masking tape, or any glue or tape that can secure the trigger in place
PROJECTILE	Anything that is slightly wider than the trigger. Reasonably dense and aerodynamic objects, such as wooden beads, pebbles, or candy, work best.

01 > Make the trigger. Cut a 6" (15 cm) piece of duct tape and split it lengthwise. Use the two pieces of tape to attach the clothespin to one end of the paint stirrer, making sure the pinching tip of the clothespin faces the middle of the stick.

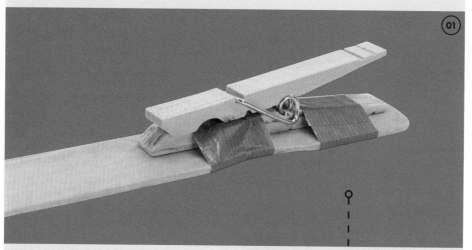

02 > Use the binder clip to clamp the rubber band to the other end of the stick. Fold the binder clip handles down. Using a binder clip allows you to quickly swap out rubber bands with different shooting power or replace broken ones.

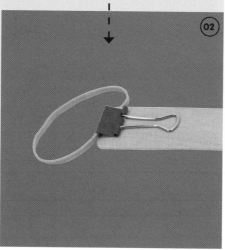

03 > Choose your projectile. For indoor shooting, try disks made by cutting a synthetic cork into quarter slices approximately ¼" to ½" (6 mm to 1.3 cm) thick. These corks are dense enough to shoot a good distance and maintain a fairly accurate trajectory but not so dense that you'll break a window. Ideally, the projectile should be slightly wider than the trigger so the rubber band can hold it in place.

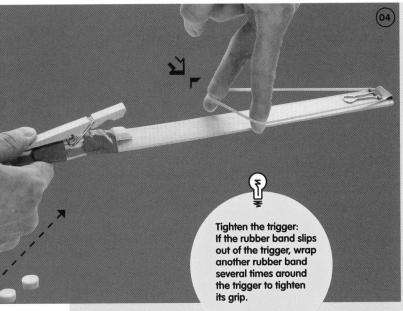

04 > You're ready to shoot! Squeeze the trigger open with one hand. Use two fingers from your other hand to stretch the rubber band into the open trigger. The rubber band should remain stretched taut when you clamp the trigger onto it.

Tighten the trigger: If the rubber band slips out of the trigger, wrap another rubber band several times around the trigger to tighten its grip.

05 > Wedge the projectile between the two sides of the rubber band directly in front of the trigger. This will prevent the projectile from falling out as you prepare your shot and ensures that the projectile receives the full force of the rubber band's elastic energy.

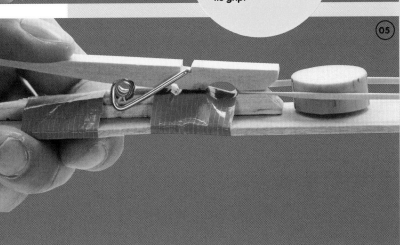

06 > Scan your surroundings for something fun to shoot at. Things that fall over, make noise, or shatter (so long as they're not of value to anyone!) are great choices. Avoid shooting point blank at hard surfaces—your projectile might ricochet back at you.

DESIGN VARIABLE

The shooter can be modified to fire *unsharpened* pencils. To prevent the pencil from veering to the side when it's fired, add a paper clip guide to the shooter. Open the paper clip into a right angle with one curved end wide enough for the pencil to pass through. Attach the paper clip to the shooter with tape.

NOW GET TINKERING

The Many-Thing Shooter can be attached to almost any flat and rigid surface. Try turning boring objects, like a binder or instruction manual, into an improvised firearm.

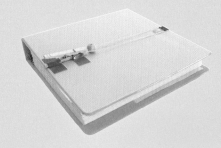

PVC
SLINGSHOT
RIFLE

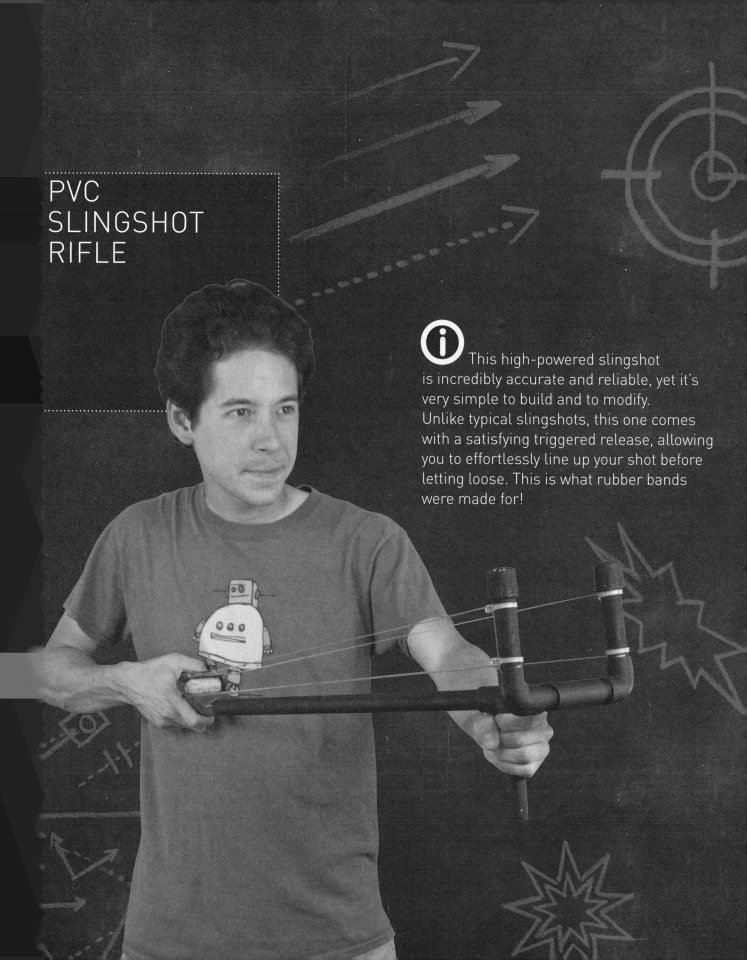

This high-powered slingshot is incredibly accurate and reliable, yet it's very simple to build and to modify. Unlike typical slingshots, this one comes with a satisfying triggered release, allowing you to effortlessly line up your shot before letting loose. This is what rubber bands were made for!

TOOLS + MATERIALS

PVC PIPE CUTTER OR HACKSAW

42" (1 M 6.5 CM) OF ½" (1.3 CM) SCHEDULE 40 PVC PIPE

TWO ½" (1.3 CM) PVC ELBOW CONNECTORS

TWO ½" (1.3 CM) PVC TEE CONNECTORS

TWO ½" (1.3 CM) PVC END CAPS (OPTIONAL)

PVC PRIMER

PVC CEMENT

BLACK SPRAY PAINT (OPTIONAL)

4 CABLE TIES

TWO 7" (18 CM) RUBBER BANDS

CARDBOARD TUBE

DUCT TAPE

LARGE BINDER CLIP

2 LARGE CRAFT STICKS

SAFETY GLASSES

MATERIAL SUBSTITUTIONS

PVC PIPE CUTTER	**Hacksaw or another type of saw**
7" (18 cm) RUBBER BANDS	**A chain of shorter rubber bands**
CARDBOARD TUBE	**Any scrap of cardboard**
LARGE CRAFT STICKS	**Any rigid piece of wood or plastic of a similar size**

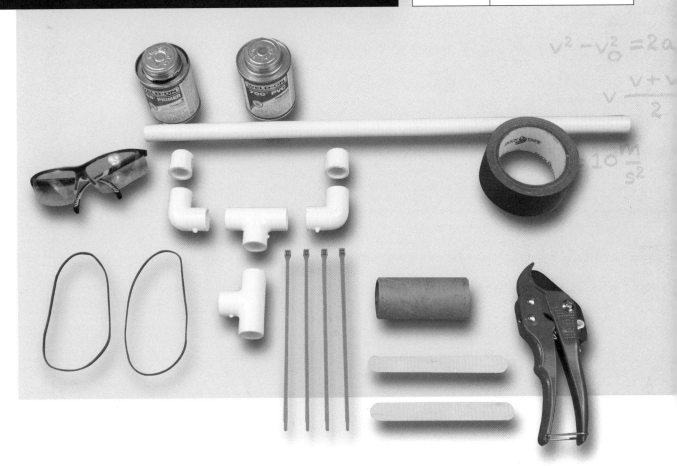

$$v^2 - v_0^2 = 2a$$

$$v \frac{v + v}{2}$$

$$10 \frac{m}{s^2}$$

01 > With the pipe cutter, cut three 2" (5 cm) lengths and three 4" (10 cm) lengths from the 42" (1 m 6.5 cm) PVC pipe. Set aside one 4" (10 cm) piece for the sling-shot's grip and the remaining 24" (61 cm) length of pipe for the handle.

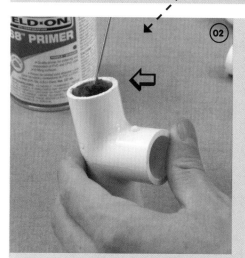

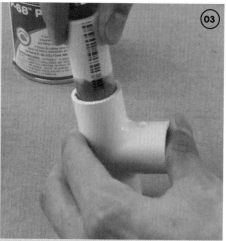

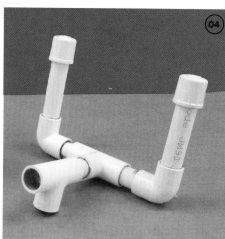

02 > Choose a well-ventilated area in which to work and protect your work surface from the primer and cement before starting. Apply primer and then cement to both ends of each piece of pipe and inside the openings of the connectors. Assemble the other short pieces of pipe, the connectors, and the end caps by simultaneously pushing and twisting the components together.

03 > Insert the 4" (10 cm) pieces of pipe into one end of each elbow connector. Use a 2" (5 cm) piece of pipe to connect each of the elbows to a tee connector. Use the remaining 2" (5 cm) piece of pipe to join the two tee connectors, turning them perpendicularly to each other. **Optional:** Top off the two 4" (10 cm) pieces of pipe with the end caps.

04 > Your slingshot will look like this. Allow the cement to set according to the manufacturer's directions.

05 > Affix the remaining 4" (10 cm) piece of pipe to the base of the slingshot as a grip. Attach the 24" (61 cm) piece of pipe to the tee connector for the handle. These two pieces of pipe don't require cementing.

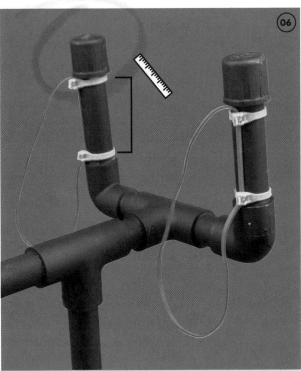

06 > Strap both ends of the rubber bands to the slingshot wye with cable ties. The cable ties should be spaced at least 3" (7.5 cm) apart. This configuration prevents the rubber bands from twisting upon release.

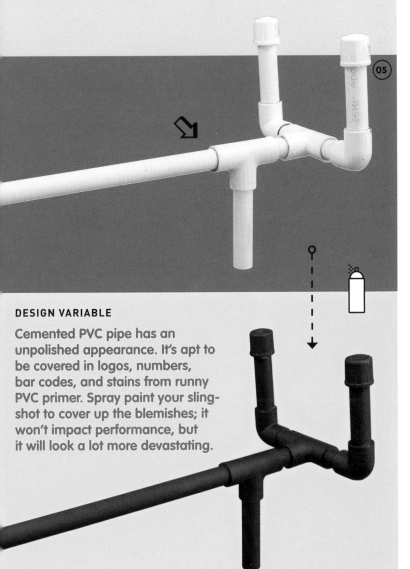

DESIGN VARIABLE

Cemented PVC pipe has an unpolished appearance. It's apt to be covered in logos, numbers, bar codes, and stains from runny PVC primer. Spray paint your slingshot to cover up the blemishes; it won't impact performance, but it will look a lot more devastating.

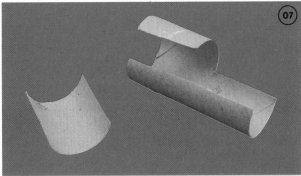

07 > Cut a 2" × 3" (5 × 7.5 cm) rectangle from the cardboard tube. This will form the slingshot's "sling." If you're using plain cardboard, roll it so it curves. You could also fashion a sling from just duct tape, but the tube offers a helpful pre-made curved shape.

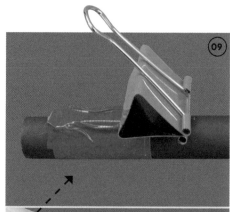

08 > Position the rubber bands around the curve of the sling. Secure the sling to the rubber bands with duct tape, and then test it by pulling it back. There should be even tension in each of the four rubber-band strands.

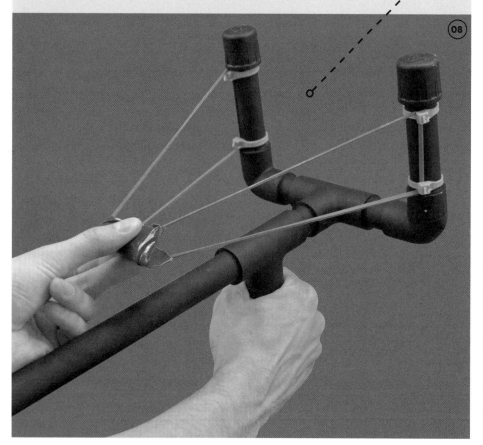

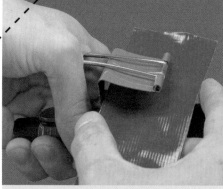

09 > Make the trigger. Attach the large binder clip to the end of the slingshot handle with duct tape. Press the binder clip open. Further secure the trigger by wrapping tape around the slingshot handle and the inside of the binder clip.

10 > Stack the two craft sticks and bind them together with duct tape. Attach the sticks to the trigger's upper binder-clip handle with tape. This will give you more leverage, allowing you to release the trigger easily.

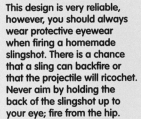
11 > Test fire! Load your projectile into the sling. Small, round, evenly weighted objects like corks and hard candy work best.

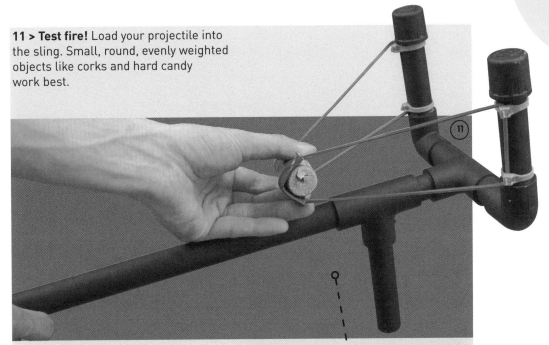

12 > Pull back on the loaded sling. Open the trigger and insert the sling. The slingshot is loaded.

13 > Fire away! Release the sling by grasping the trigger handle.

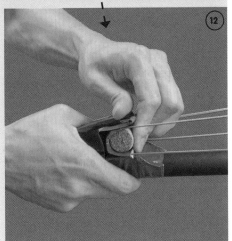

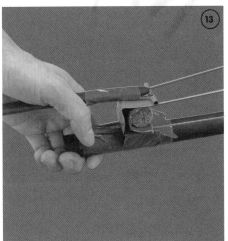

This project requires some determination to complete, but the payoff is worth it. The crossbow makes a satisfying *snap* when the trigger is pulled, and it can launch bolts more than 100' (30 m)! You can power this crossbow with a piece of string or with a rubber band.

CROSSBOW

01 > Begin by preparing the paint stirrers as shown here. For the first, drill a ⅛" (3 mm) hole near each end of the stick. For the second and third, drill a ⅛" (3 mm) hole that is 5¼" (13.5 cm) from one end of each stick.

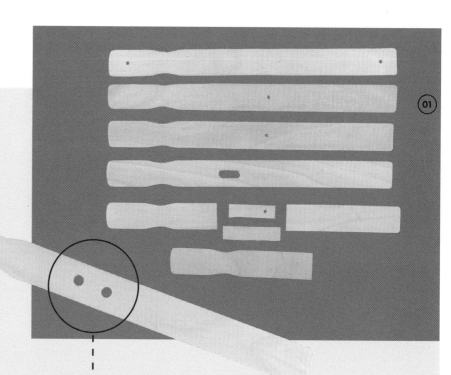

02 > For the fourth paint stirrer, using the ⅜" (1 cm) bit, drill a hole 4¾" (12 cm) from one end. Lift the drill, move it toward the center of the stick, and drill a second hole ½" (1.3 cm) from the first.

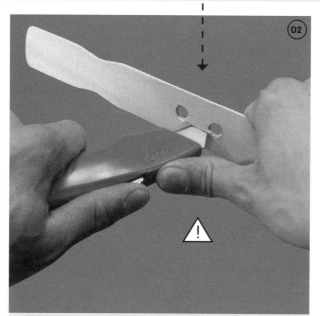

03 > Using the utility knife, carve out the section between the two holes created in step 2. Carefully cut along the grain of the wood by slowly pushing the knife away from your body and hands.

04 > For the fifth paint stirrer, use the utility knife to score and break the stirrer into two 4¾" (12 cm) pieces and one 2" (5 cm) piece. Scoring the wood with the blade as described in step 3, split the 2" (5 cm) piece in half lengthwise. Trim ¼" (6 mm) off one of the 2" (5 cm) pieces and drill a ⅛" (3 mm) hole near the end of it.

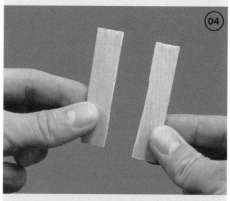

05 > Cut the sixth paint stirrer in half, crosswise. You'll only need one of the halves.

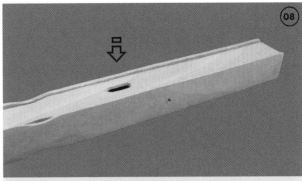

06 > Begin assembling the shaft of the crossbow like a box. Line up the paint stirrer from step 3 with one of the stirrers from step 1 with a hole 5¼" (13.5 cm) from the end. Take careful note of the orientation of the holes in the photo.

07 > Use hot glue to attach the edges of the paint stirrers at right angles. If the glue dries too quickly when you run a thin line of it along the edge, use several beads of glue, instead.

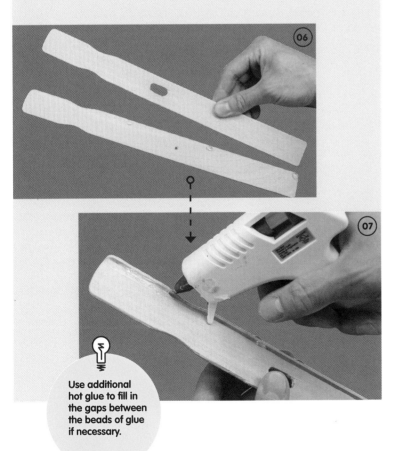

Use additional hot glue to fill in the gaps between the beads of glue if necessary.

08 > Glue the second stirrer with a hole 5¼" (13.5 cm) from the end opposite the first. Make sure that the two ⅛" (3 cm) holes line up.

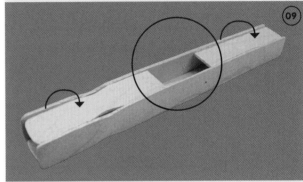

09 > Turn the box over and glue the two 4¾" (12 cm) stirrer pieces to the underside of the shaft. The 2" (5 cm) gap in the center will be where the trigger goes.

✓

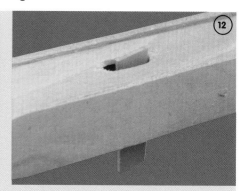

10 > **Make a trigger** with the two small pieces of wood from step 4. Position the piece with the ⅛" (3 mm) hole at a right angle to the other, ½" (1.3 cm) down from the top. Hot glue the two pieces together.

11 > Set the trigger into place. Choose the thickest skewer and thread it through the holes in the crossbow shaft and trigger. This design relies on the friction between the skewer and the ⅛" (3 mm) holes to hold it in place. The ½" (1.3 cm) of the trigger's crosspiece should poke through the cut-out slot.

12 > This is how the trigger should look, and it should swing up and down. If it doesn't quite fit, then you may need to drill or carve out a larger hole or drill new holes for the trigger hinge. When you have it working, trim off the extra lengths of skewer and save the scraps.

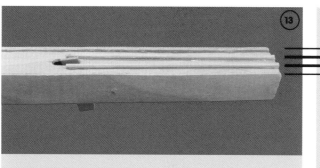

13 > **Create the guides for the bolt.**
Cut the skewer scrap and the second skewer to 6¾" (17 cm) lengths. Glue the two skewer pieces parallel to each other with about ⅜" (1 cm) between them. The ends of the skewers should line up with the front of the crossbow shaft and the middle of the trigger head. Make sure the ends of the skewer nearest the trigger have a clean and flat cut, or the string might not latch on correctly.

14 > Center and glue the paint stirrer with the ⅛" (3 mm) holes drilled at each end to the front end of the crossbow shaft.

15 > Thread the string through the ⅛" (3 mm) holes and knot the ends. To make the knot tying easier, an 18" (45.5 cm) length of string is recommended, but ultimately the string should be 12" (30.5 cm) long from one hole to the other. The bow should start to bend as you pull the string back a few inches.

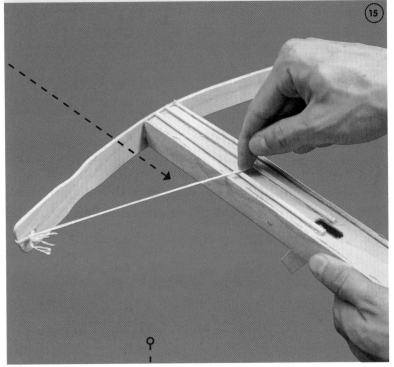

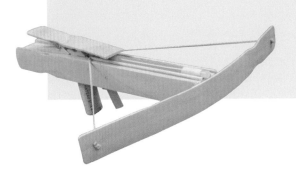

MATERIAL SUBSTITUTION

Here is a crossbow outfitted with a rubber band, which may be the way you'd like to go. Getting just the right string length is a trial-and-error process. Rubber bands are easier to load and easier to calibrate to achieve the right amount of energy, but they may not provide as much force as a taut string.

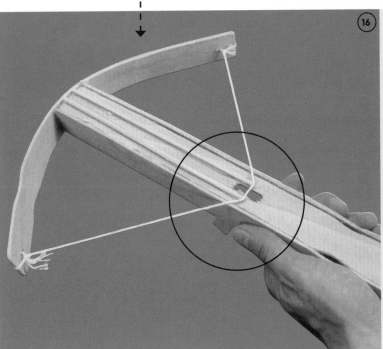

16 > Give the string a test by pulling it back and slipping it over the ends of the skewers. (If this is difficult to do with your fingers, use the pencil's eraser to push the string into place.)

You may need to calibrate the tension of the string to achieve the most force. The string should be a little loose when under no tension, but very taut when loaded. If you are unable to load the string at all, then it needs to be a little longer. Trim the ends of the string with scissors when complete.

Note: If the string is slipping off of the skewers, use a utility knife to cut each skewer end, either flat or slightly indented.

17 > Glue the clothespin directly behind the cut-out slot.

You can use the clothespin as a simplified trigger for the rubber band–based design.

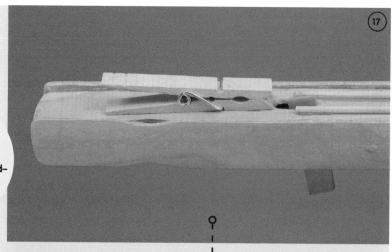

18 > Glue the half piece of paint stirrer from step 5 onto the top of the clothespin. The lower end of the stirrer should be between ¼" and ⅜" (6 mm and 1 cm) from the top of the shaft. This will be used to hold bolts in place.

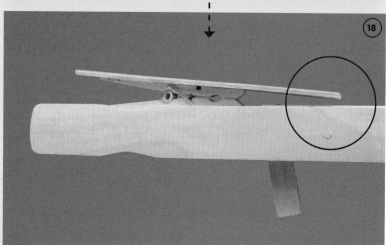

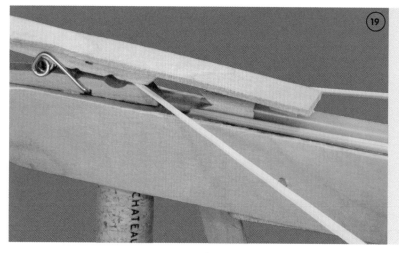

19 > Test the trigger. Hot glue the cork approximately 1½" (4 cm) behind the trigger. Pulling on the trigger should push the string off the ends of the skewers. If the string gets caught on one skewer, double-check to make sure that the trigger is centered exactly between the skewers. Also check to make sure that the ends of the skewers are about the same diameter and flat.

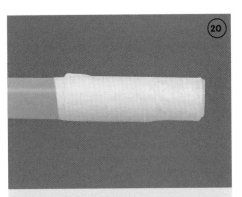

Create Crossbow Bolts

Now it's time to create the crossbow bolts. There are many options for creating ammunition: pens, pencils, dowels, and even hard candy will work. These bolts, made from drinking straws, are designed for distance.

20 > Cut a thick drinking straw to 6" (15 cm) in length. Insert something dense in the tip, like a 2" (5 cm) piece of a hot glue stick. This piece of glue is called a leading weight. (See the notes on leading weights on page 34.)

Tape the glue stick in place and wrap tape around the other end of the straw.

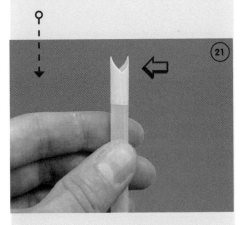

21 > Cut a nock into the back of the straw by pinching the tip until it's flat, then cutting off the corners. A nock will ensure that the bolt engages the string and fires consistently.

22 > Get ready to fire! Load the string behind the skewers on the crossbow. Place your bolt directly in front of the trigger hole, but don't cover it. Close the bolt holder on top of the bolt to keep it in place while you prepare your shot.

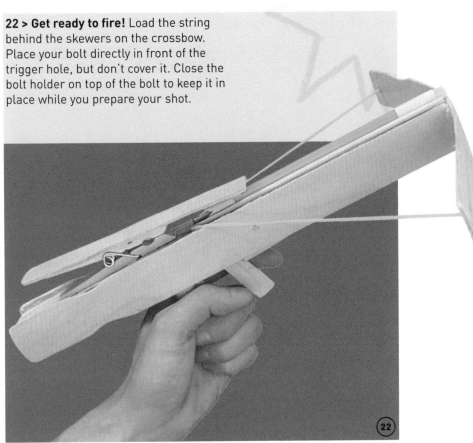

DESIGN VARIABLES

This is just one example of a crossbow, but you can scale the proportions of the crossbow frame to be larger or smaller.

Do a quick internet search for "crossbow trigger diagram" to find more complex ways of releasing the string.

Add a thumbtack to the tip of your bolt and set up some cardboard targets.

Create a compartment for holding your ammunition.

BOW AND ARROW

(i) Branches, sticks, string, rubber bands, pens, rulers, and bamboo skewers can all be used to craft a bow and arrow. This design is one way to do it, and it illustrates the important principles. Like all the projects in this book, it's built to be as fun and effective as it looks.

TOOLS + MATERIALS

2 YARDSTICKS (OR METER STICKS)
DUCT TAPE
UTILITY KNIFE
PLIERS (OPTIONAL)
STRING
DOWEL, 24" (61 CM) LONG
PENCIL SHARPENER (OPTIONAL)
SPOOL OF BENDABLE, NON-ALUMINUM WIRE, AT LEAST 18 GAUGE

MATERIAL SUBSTITUTIONS

YARDSTICKS	Tape several paint stirrers together in an overlapping pattern; any strong yet flexible material that can be cut to the dimensions of a yardstick
STRING	A long chain of linked rubber bands
DOWEL	Bamboo gardening stakes

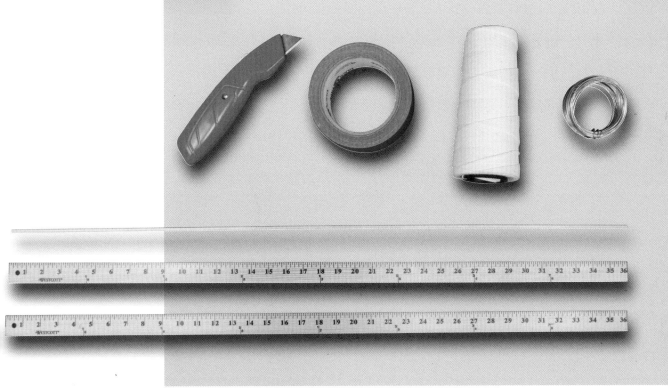

01 > Stack one yardstick (meter ruler) on top of the other and tightly wrap the ends together with duct tape. If your yardsticks have a hole at one end, make sure that both holes line up.

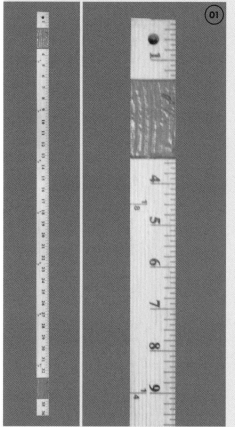

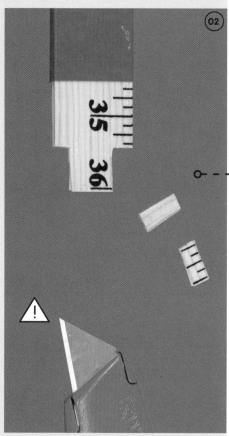

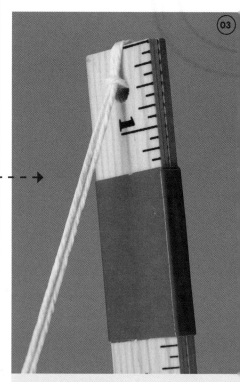

02 > Cut a ¼" × ½" (6 mm × 1.3 cm) piece off each corner of the yardstick. Use the utility knife to score the yardstick several times and then break the piece off. Because the grain of the wood runs lengthwise along the yardstick, you should have a clean break. Use pliers to snap off the corners if you have trouble breaking the wood.

03 > Cut a 6' 8" (2 m 20.5 cm) length of string. Fold the string in half and attach it to one end of the bow with a hitch knot. If your yardsticks have a hole at one end, thread the folded end of the string through the hole.

04 > String the bow by bending it significantly while holding the loose ends of the string in your hand. Wrap the string around the second end of the bow and knot it.

05 > Cut a 1" (2.5 cm) piece from the dowel. Tape the 1" (2.5 cm) piece of dowel so that it lines up with the very center of the bow. (If you are using yardsticks, this is at the 18" mark [50 cm] on a meter ruler.) This will serve as the arrow rest.

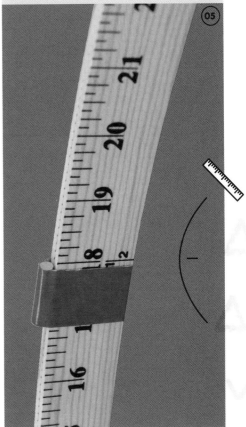

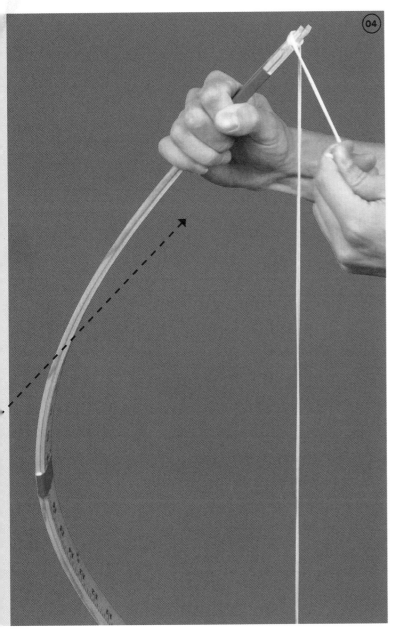

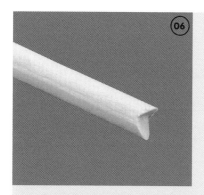

07 > Use the utility knife to whittle the other end of the dowel to a point. Alternatively, use a pencil sharpener for a more precise point.

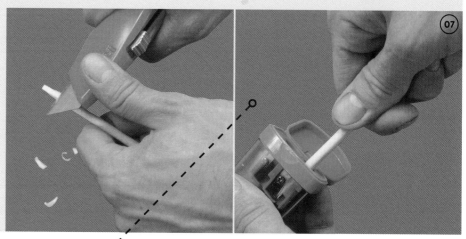

06 > Now turn the long piece of dowel into an arrow. Use the utility knife to cut a V-shaped nock at one end of the dowel. The nock will prevent the arrow from slipping off the bow string when you shoot.

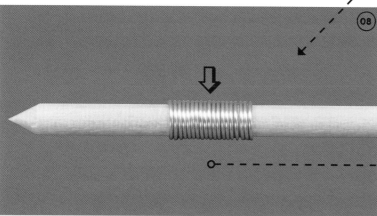

08 > Wrap fine wire around the dowel near the arrow's tip to create a 1" (2.5 cm)-wide band. Adding this leading weight will allow the arrow to fly straight.

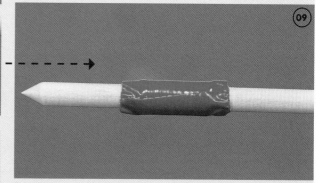

09 > Wrap the wire with duct tape to secure it. You're ready to fire!

THE IMPORTANCE OF LEADING WEIGHTS

Imagine trying to throw a long strip of paper. It won't go very far and it definitely won't go straight. Now imagine attaching a rock to one end of the paper and throwing it again. The momentum of the rock will carry the paper through the air and the paper will trail behind it in a straight line. This same idea applies to long projectiles—such as arrows. They need weight at the tip in order to work.

10 > Nock the arrow onto the string and rest the shaft on the arrow rest. Contrary to the techniques of real archery, you will pinch the nocked dowel and draw it back until only the very tip of the arrow is in front of the bow. Aim safely, and let go! Sheets of cardboard make great targets.

WAIT, WHERE'S THE FLETCHING?

Real arrows use fletching (fins or feathers) in addition to a leading weight to stabilize the arrow's flight. Arrows made from household materials sometimes don't. Here's why:

Real arrows are flexible. If you watch an arrow being shot in slow motion, it will bend significantly under the force of the released bowstring. This bending allows the arrow to effectively curve around the bow before straightening out in midflight. This is important!

If the arrow did not bend, the fletching would collide with the bow and be ruined. That is exactly what will happen if you add fletching to this inflexible arrow. If you make an exceptionally strong bow, or you use a more flexible material for the arrow, your arrows will benefit from fletching. Want to know more? Do an internet search for "the archer's paradox."

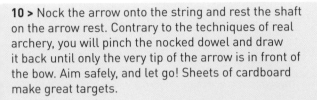

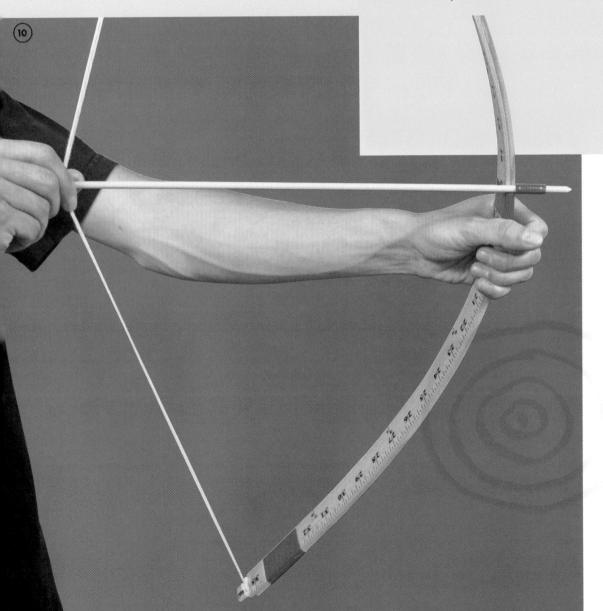

MINI SIEGE
ENGINES

$$v \frac{v + v_0}{2}$$

$$= 10 \frac{m}{s^2}$$

PYRAMID CATAPULT

ENHANCED MOUSETRAP CATAPULT

FLOATING ARM TREBUCHET

DA VINCI CATAPULT

Catapults don't have to be complicated to impress. This design is simple, robust, and surprisingly effective.

PYRAMID
CATAPULT

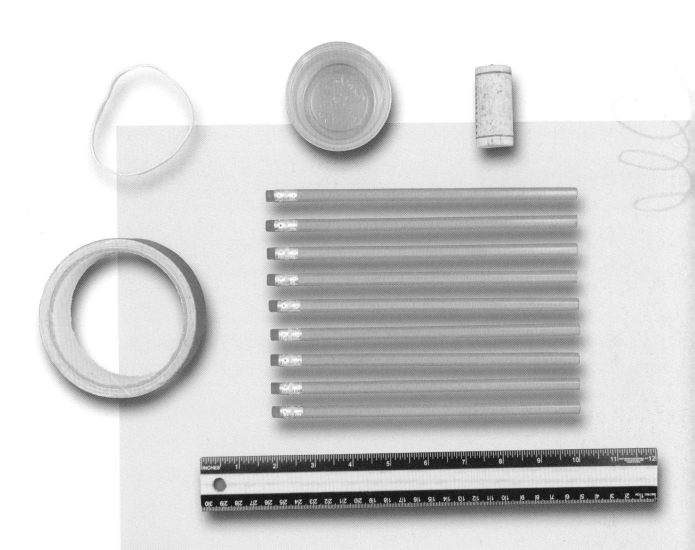

MATERIAL SUBSTITUTIONS	
PENCILS	Craft sticks, dowels, or other implements with a similar shape and strength
DUCT TAPE	Masking tape, electrical tape
WOODEN RULER	Paint stirrer, craft sticks glued together
CORK	Small, dense objects

01 > Position two pencils on top of a 3" to 4" (7.5 to 10 cm) piece of duct tape. Do your best to arrange the pencils in an equilateral triangle shape. Wrap tape tightly around the ends of the pencils where they meet.

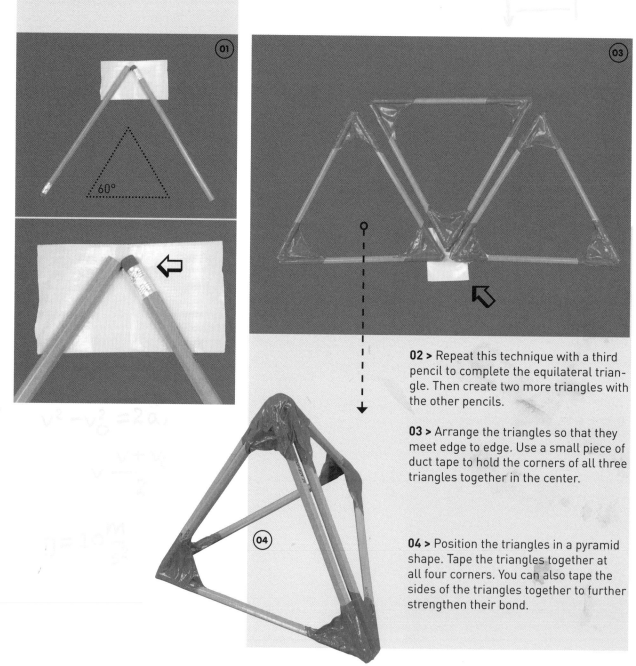

02 > Repeat this technique with a third pencil to complete the equilateral triangle. Then create two more triangles with the other pencils.

03 > Arrange the triangles so that they meet edge to edge. Use a small piece of duct tape to hold the corners of all three triangles together in the center.

04 > Position the triangles in a pyramid shape. Tape the triangles together at all four corners. You can also tape the sides of the triangles together to further strengthen their bond.

DESIGN VARIABLES

The size of this design is highly variable. You can make miniature catapults or create a giant one. As a rule of thumb, you will need stronger materials for larger designs.

What you choose to launch is up to you. Corks have a good balance of density, aerodynamics, and safety.

05 > Position one end of the ruler on a 3" to 4" (7.5 to 10 cm) piece of duct tape. Leave about half of the length of the tape free.

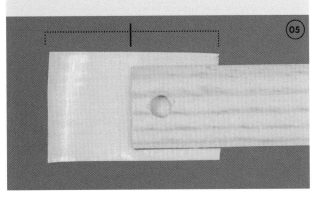

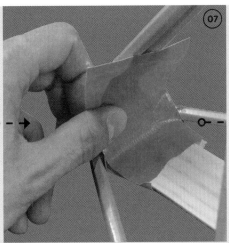

06 > Position the pyramid and ruler as shown, with one corner of the pyramid on the exposed tape. Fold the tape around both sides of the corner.

07 > Use another piece of duct tape to attach the upper side of the ruler to the pyramid. This creates a duct-tape hinge.

08 > Finish securing the hinge by folding the duct tape around the corner of the pyramid.

ADD EXTRA POWER!

Use multiple rubber bands to give your pyramid catapult extra *umph*. But if you reach a point where you are unable to draw the catapult arm all the way back or your frame starts to fall apart, you've added one rubber band too many.

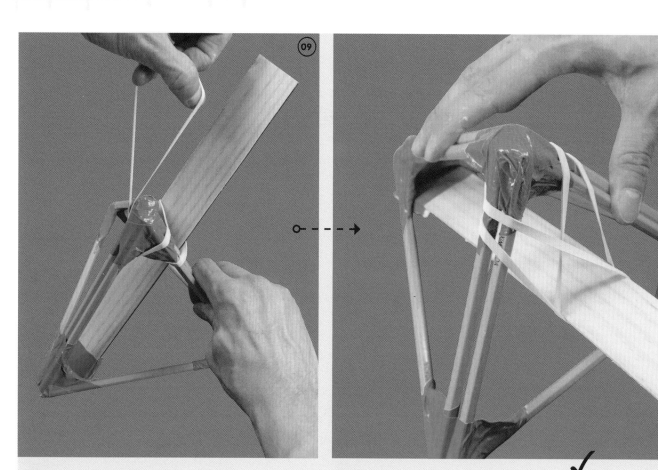

09 > Time for the rubber band. Loop the thick rubber band over the ruler, pull it through the pyramid, and then stretch it over the ruler again.

10 > The rubber band should look like this when it is attached correctly. This configuration ensures that the force of the rubber band is equally distributed.

11 > Center a 5" (12.5 cm) piece of duct tape on the back of the ruler about 1" (2.5 cm) from the end.

12 > Attach the cup by pressing the tape tightly against the sides.

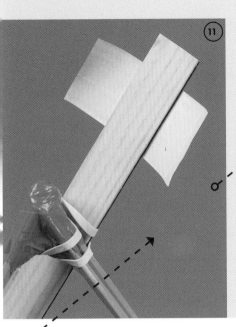

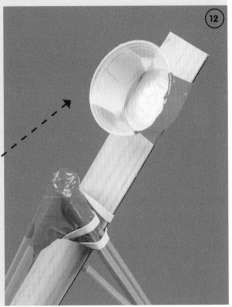

13 > To fire, take the catapult outside or into a room with high ceilings. Load your projectile into the cup. Hold down the front corner of the pyramid. Place your thumb on the exposed inch of ruler behind the cup. Pull down on the ruler as far as it will go and release it quickly.

⚠ Make sure there's nothing breakable around. You'll be surprised by how far and fast this simple design can throw!

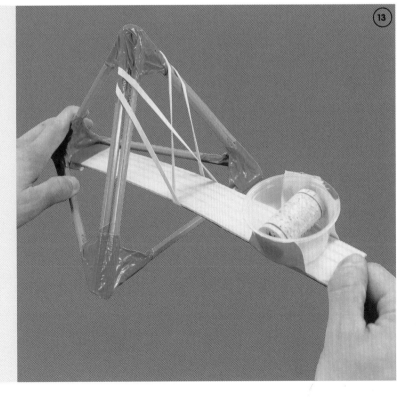

ENHANCED
MOUSETRAP
CATAPULT

ⓘ If you do an internet search for "mousetrap catapult," you'll find dozens of designs. Sometimes the engineering is clever, but mousetrap catapults often fail to impress for two reasons: the mousetrap's torsion spring has a limited and unchangeable amount of potential energy, and the angle of the catapult is static. The Enhanced Mousetrap Catapult utilizes the torsion spring, but also increases the amount of energy with extra rubber bands. Furthermore, this design includes a simple mechanism for adjusting the trajectory, so you can lob projectiles over high walls or fire straight into them!

SIX 12" (30.5 CM) PAINT STIRRERS

MOUSETRAP

NEEDLE NOSE PLIERS

DUCT TAPE

DRILL WITH 1⁄8" (3 MM) BIT

RULER

MARKER OR PEN

RUBBER BANDS

UTILITY KNIFE

HOT GLUE GUN

1⁄8" (3 MM)-THICK BAMBOO SKEWER

SMALL CUP

MATERIAL SUBSTITUTIONS

PAINT STIRRERS	Craft sticks, wood rulers, or wood trim from a hardware store
UTILITY KNIFE	Hacksaw or other wood-cutting tool

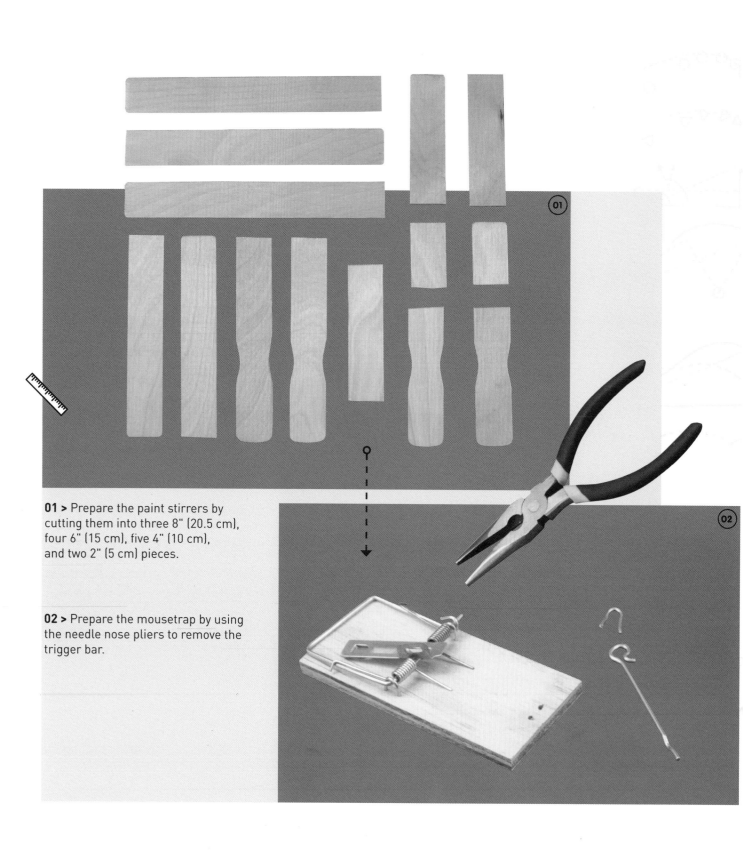

01 > Prepare the paint stirrers by cutting them into three 8" (20.5 cm), four 6" (15 cm), five 4" (10 cm), and two 2" (5 cm) pieces.

02 > Prepare the mousetrap by using the needle nose pliers to remove the trigger bar.

03 > Hot glue the four 6" (15 cm) pieces of paint stirrer into two A-frame shapes. The widest part of the opening between the legs of the A-frame should measure 4" (10 cm).

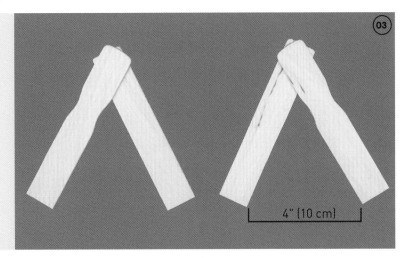

4" (10 cm)

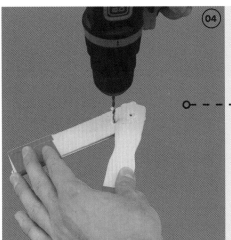

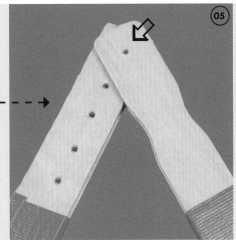

04 > Stack the two A-frames and wrap the ends in duct tape to hold them together. Drill a series of holes along one leg of the A-frames, starting at the apex.

05 > Aim for five holes spaced about ½" (1.3 cm) apart.

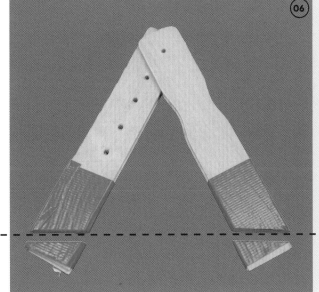

06 > Place a ruler across the bottom of the A-frames and use a marker or pen to draw a line across the legs. Trim the bottom of each leg. Using a utility knife, score the wood along the marked line several times then snap off the end with your fingers or the pliers.

07 > Create a rectangular base for the catapult. Lay out two parallel 8" (20.5 cm) pieces of paint stirrer. Hot glue three 4" (10 cm) pieces of paint stirrer across them as shown. The two 4" (10 cm) pieces that are closest to each other should be 2" (5 cm) apart.

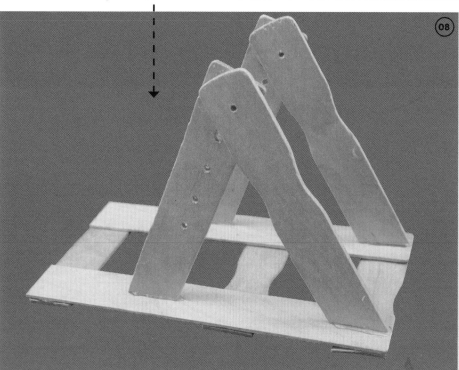

08 > Remove the tape and separate the A-frames. Glue the A-frames to the base, making sure the drilled holes are aligned. Note that the frames are positioned toward the front of the catapult where the mousetrap will be.

09 > Sandwich the mousetrap bar between the two 2" (5 cm) pieces of paint stirrer and use plenty of hot glue to attach the three layers. Allow the glue to set completely.

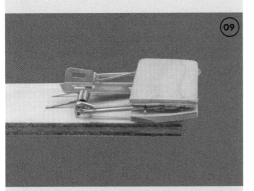

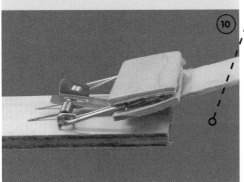

10 > Glue the remaining 8" (20.5 cm) piece of paint stirrer to the under part of the sandwich you just created, as shown.

11 > Glue the mousetrap assembly to the front of the catapult with the catapult arm tilted forward.

12 > Lift the catapult arm and insert the bamboo skewer through a pair of drilled holes in the A-frames. This will keep the arm in place while you finish the catapult.

13 > Glue the two remaining 4" (10 cm) pieces of paint stirrer to the front of the A-frame for stability.

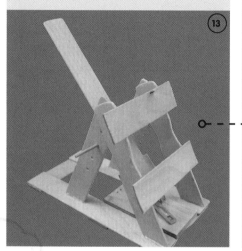

14 > Place a piece of duct tape on the back of the catapult arm, about 1" (2.5 cm) from the end.

15 > Position the cup over the duct tape and wrap the tape tightly around the sides of the cup.

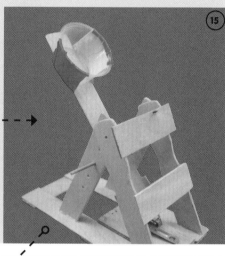

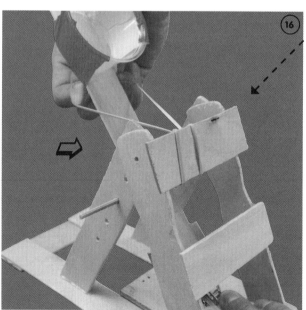

16 > Use a hitch knot (slipping one end through the other) to attach a rubber band to the frame brace then loop the rubber band over the catapult arm.

17 > Attach a second (or a third or fourth) rubber band for added power!

18 > The trajectory of the projectile is determined by the angle of the catapult arm when it's at rest. You can adjust the catapult's trajectory by removing and reinserting the skewer at different levels. As a rule of thumb, the projectile will launch in the same direction that the cup is facing. In the first picture, the catapult has a very high trajectory, and in the second, it has a straight trajectory.

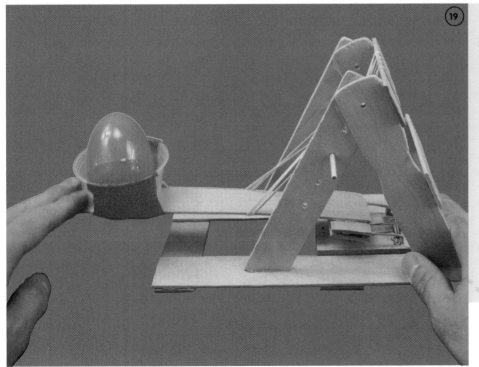

19 > Load your favorite projectile. Hold down the front of the catapult with one hand and draw back the arm with your other hand. Release by allowing your fingertips to slip off the catapult arm.

The floating arm trebuchet is a modern take on an ancient war machine. This variation features a drop channel, which allows the counterweight to fall straight down rather than swing with the arm. This is a more efficient way to transfer potential energy into kinetic energy on a small scale. A pair of wheels allows the arm to roll along glide rails as the counterweight falls. This design is more complicated to build than a traditional trebuchet, but the launching mechanism is more effective and much more gratifying to behold!

FLOATING ARM TREBUCHET

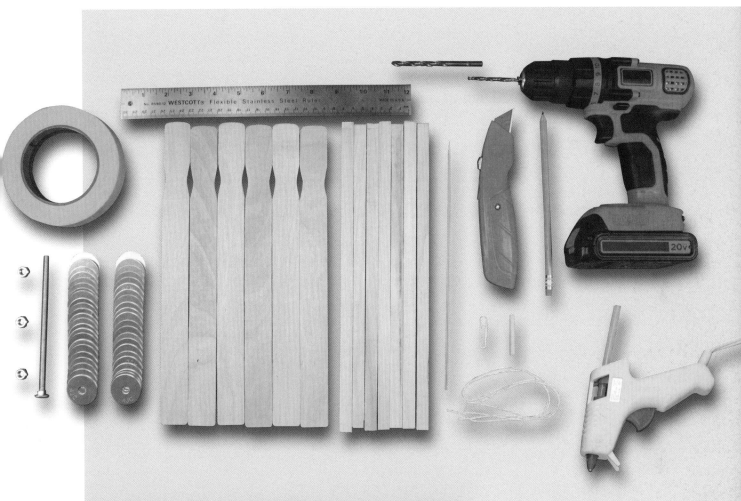

TOOLS + MATERIALS

SIX 12" (30.5 CM) PAINT STIRRERS

UTILITY KNIFE

RULER

HOT GLUE GUN

DRILL WITH ¼" AND ⅛" (6 MM AND 3 MM) BITS

SEVEN 12" (30.5 CM) SQUARE DOWELS, ½" (1.3 CM) WIDE*

¼" (6 MM) BOLT, 5" (12.5 CM) LONG

SEVENTY ¼" (6 MM) FENDER WASHERS

THREE ¼" (6 MM) HEX NUTS

¼" (6 MM) WOODEN DOWEL

MASKING TAPE

⅛" (3 MM)-THICK BAMBOO SKEWER

AT LEAST 30" (76 CM) OF STRING

PAPER CLIP

MATERIAL SUBSTITUTIONS	
PAINT STIRRER	Ruler, wood shim, craft sticks, or another flat, rigid strip of wood
MASKING TAPE	Duct tape

*Square dowels must be perfectly straight or the trebuchet may not operate smoothly.

01 > Prepare the pieces:

Cut two 12" (30.5 cm) paint stirrers into four 6" (15 cm) pieces.

Split two of the 6" (15 cm) pieces down the middle.

Cut the bottom of the split pieces at approximately a 45-degree angle.

Cut 2 paint stirrers into four 5" (12.5 cm) pieces, leaving two 2" (5 cm) pieces.

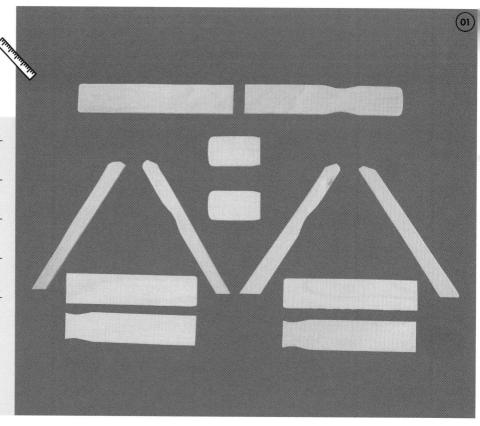

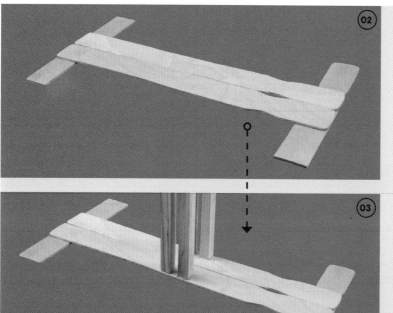

02 > Form the base using two whole paint stirrers and the two 6" (15 cm) pieces. Center and hot glue the ends of the long pieces to the shorter cross pieces.

03 > Create the drop channel. Hot glue four square dowels upright at the center of the base. The gap between the pair of dowels on each side of the base is ¼" (6 mm) wide. The space between the dowel pairs across the base is 1½" (4 cm).

04 > Use the 5" (12.5 cm) pieces of paint stirrer and the four 6" (15 cm) square dowel pieces to form the glide rails.

05 > The glide rails do not cross into the drop channel.

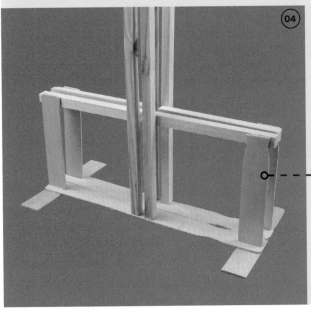

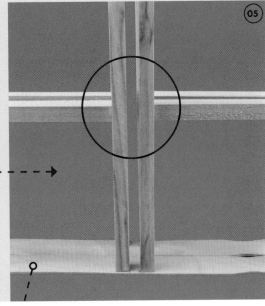

06 > Use the 2" (5 cm) pieces of paint stirrer to add structural support to the glide rails and to help maintain a uniform gap.

07 > Glue the split paint stirrer pieces at an angle to support the glide rails and drop channel. The tops of these supports have been trimmed for a cleaner look.

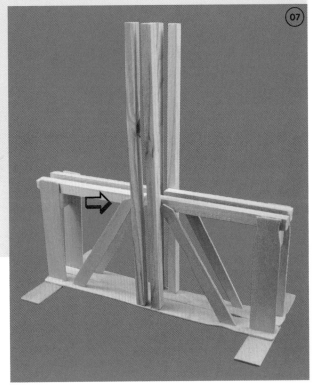

08 > Get ready to assemble the counterweight and trebuchet arm.
You'll need all the hex nuts and all but 10 of the washers. The washers and nuts will be assembled onto the bolt in the order shown.

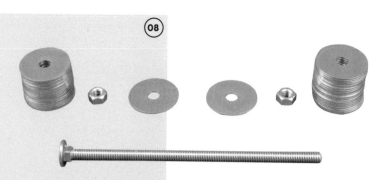

09 > Drill four ¼" (6 mm) holes into a square dowel. One hole should be near the end of the dowel, and the other three are spaced 1" (2.5 cm) apart, starting 2" (5 cm) from the first hole.

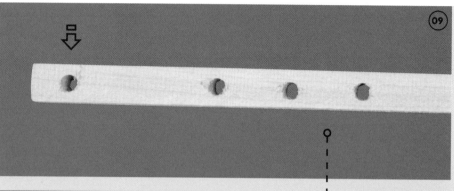

10 > Assemble the metal components with the square dowel in the center. Insert a 2" (5 cm) piece of ¼" (6 mm) round dowel into one of the remaining holes.

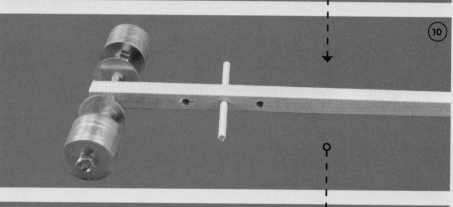

11 > Wrap layers of masking tape around the round dowel until the thickness is slightly larger than the center of the washers. Snugly fit 5 washers onto each end of the dowel. The dowel should spin freely, but the washers should remain in place. Cut off any excess dowel to prevent it from getting caught on the drop channel beams. This will be the roller that travels over the glide rails.

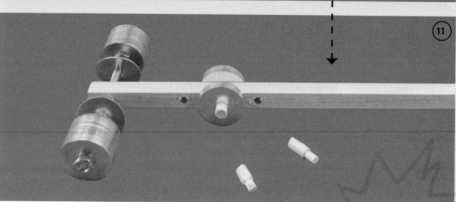

12 > Set the counterweighted arm into the drop channel. Give it a quick test by pulling the arm back and letting it go. The arm should smoothly swing forward as the counterweight falls.

If the arm is colliding with the glide rails, carefully detach the drop channel dowels from the base and reglue them slightly farther apart.

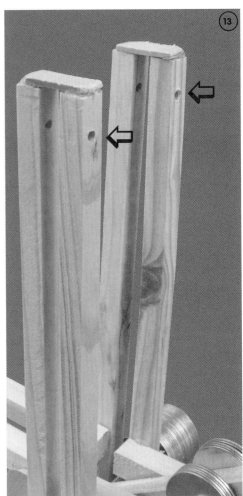

13 > Once the arm and counterweight are swinging smoothly, use two scraps of wood to close off the top of the drop channel. Drill two sets of holes using the ⅛" (3 mm) bit for the trigger pins.

MAXIMIZING PERFORMANCE

There are several subtle but key variables that influence how well the trebuchet performs. The rollers in the arm can be positioned into different holes, thus altering the position of the fulcrum. The hole nearest the end of the arm maximizes the speed of the projectile, but it may not perform as well with heavier projectiles. As discussed below, the angle of the paper clip is a key variable, as well as the type of projectile, and the length of string to which it's attached.

14 > Create the trigger by tying and taping one 30" (76 cm) piece of string to two 2" (5 cm) pieces of bamboo skewer.

15 > Set the trigger by raising the counterweight above the holes and inserting the pins. The pins should fit in the holes loosely. The goal is to use the weight of the counterweight (rather than friction) to hold the pins in place.

16 > Straighten out a paper clip and then fold it in half. Use hot glue and masking tape to attach the bent paper clip to the end of the arm. Bend the rounded end of the paper clip upward slightly. This is where the projectile will be attached.

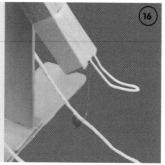

EXPERIMENT WITH THE PAPER CLIP ANGLE

The angle of the bend in the paper clip will determine the timing of the projectile's release. A straighter paper clip will result in a quicker release and a higher trajectory. A paper clip with greater curve will give a delayed release and a straighter trajectory. There's no "best" angle—it depends on the trajectory you want to achieve.

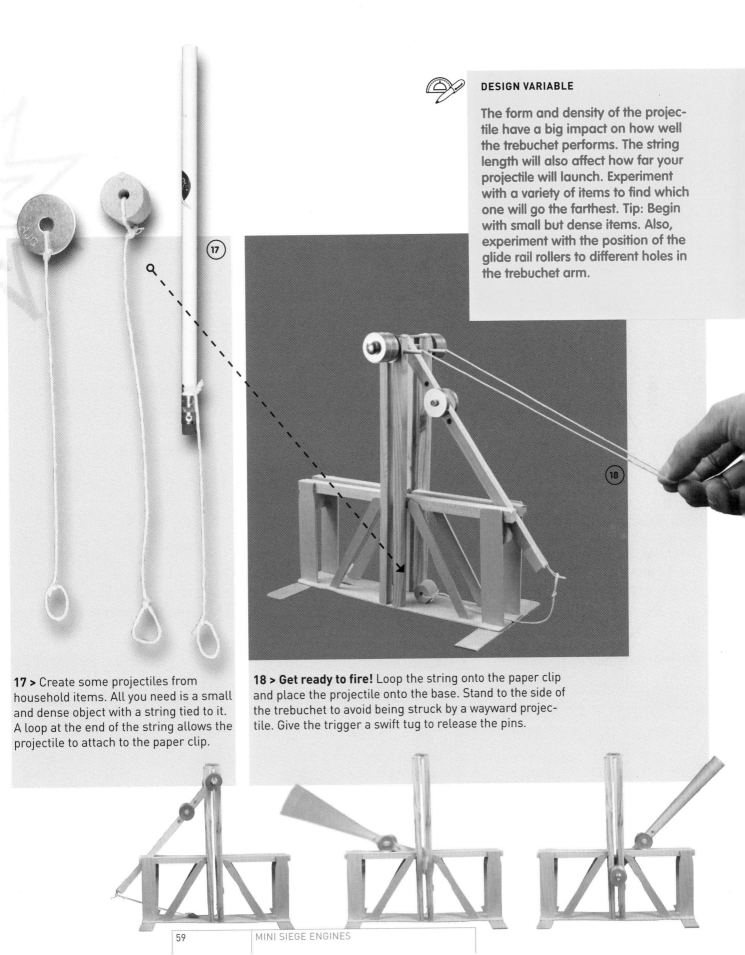

DESIGN VARIABLE

The form and density of the projectile have a big impact on how well the trebuchet performs. The string length will also affect how far your projectile will launch. Experiment with a variety of items to find which one will go the farthest. Tip: Begin with small but dense items. Also, experiment with the position of the glide rail rollers to different holes in the trebuchet arm.

17 > Create some projectiles from household items. All you need is a small and dense object with a string tied to it. A loop at the end of the string allows the projectile to attach to the paper clip.

18 > Get ready to fire! Loop the string onto the paper clip and place the projectile onto the base. Stand to the side of the trebuchet to avoid being struck by a wayward projectile. Give the trigger a swift tug to release the pins.

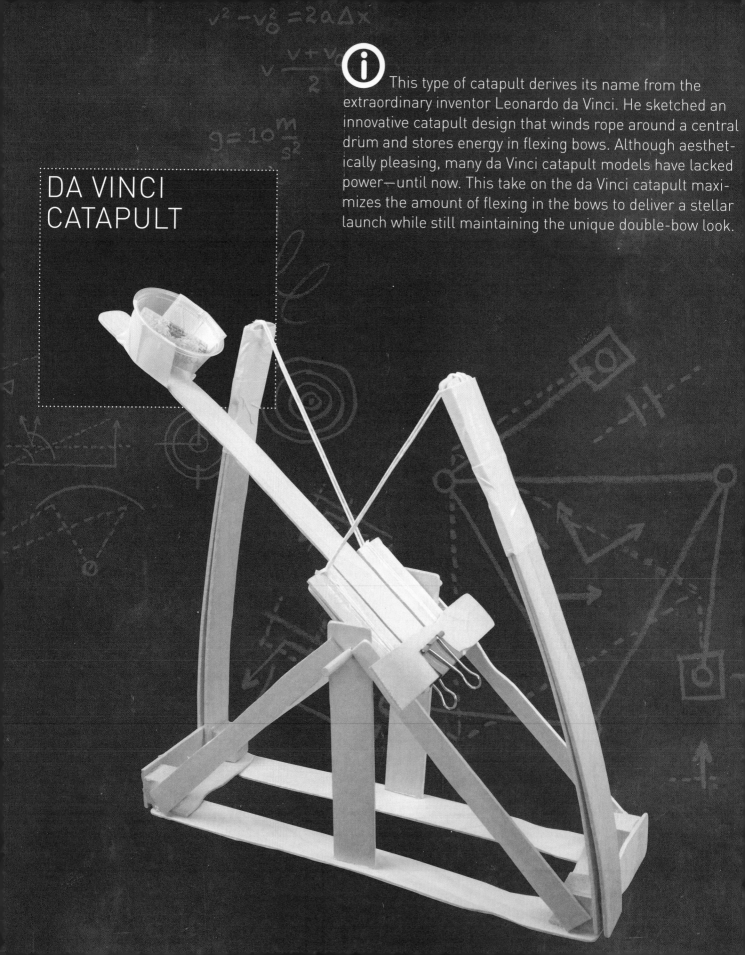

DA VINCI CATAPULT

ⓘ This type of catapult derives its name from the extraordinary inventor Leonardo da Vinci. He sketched an innovative catapult design that winds rope around a central drum and stores energy in flexing bows. Although aesthetically pleasing, many da Vinci catapult models have lacked power—until now. This take on the da Vinci catapult maximizes the amount of flexing in the bows to deliver a stellar launch while still maintaining the unique double-bow look.

TOOLS + MATERIALS

THIRTEEN 12" (30.5 CM) PAINT STIRRERS

MEASURING TAPE

UTILITY KNIFE

DRILL WITH ¼" (6 MM) BIT

¾" (2 CM) SQUARE DOWEL

HACKSAW

SANDPAPER (OPTIONAL)

HOT GLUE GUN

PENCIL

¼" (6 MM) ROUND DOWEL

STRING

DUCT TAPE

BINDER CLIP

SMALL CUP

PROJECTILE OF YOUR CHOICE

MATERIAL SUBSTITUTIONS	
PAINT STIRRER	For the bow, any semi-flexible, flat material such as a wood ruler. The other paint stirrers can be replaced with any rigid material, such as craft sticks or wood shims.
HACKSAW	Any tool that can cut the ¾" (2 cm) dowel

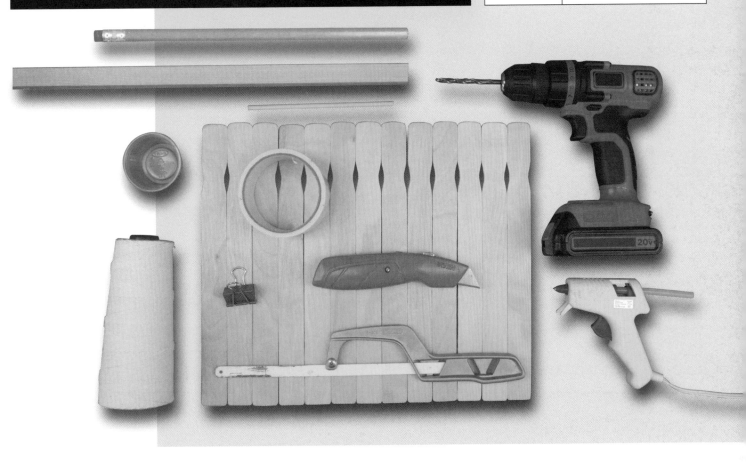

01 > Select two paint stirrers with very straight wood grain. Cut the two stirrers to 7½ (19 cm) and save the remaining pieces. With the utility knife, score the 7½" (19 cm) pieces in half lengthwise and then break apart. If the grain of the wood is straight, they should split evenly.

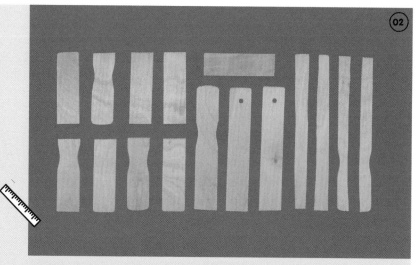

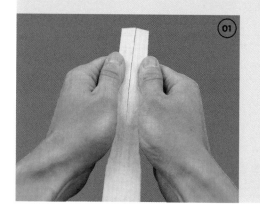

02 > Prepare the remaining pieces of paint stirrers. Cut nine 3½" (9 cm) pieces. Cut three 6" (15 cm) pieces. Drill a ¼" (6 mm) hole near the end of two 6" (15 cm) pieces.

03 > Form a rectangular base, using two 3½" (9 cm) pieces of paint stirrer and two whole stirrers. Hot glue the pieces together at the corners.

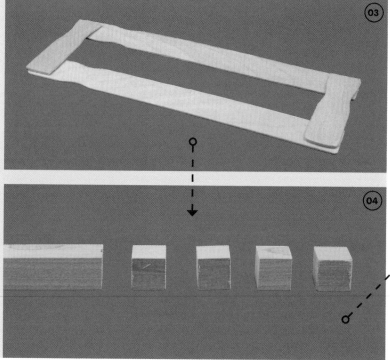

04 > Cut four ¾" (2 cm) cubes with the hacksaw. **Optional:** Clean up the edges of the cubes with sandpaper or by scraping them against another abrasive surface, such as concrete.

07 > Position a 3½" (9 cm) piece of paint stirrer horizontally from the bottom of each bow to the next cube. Hot glue the pieces into place at each end. Line up a second bow with each of the first and hot glue them into place at the base only.

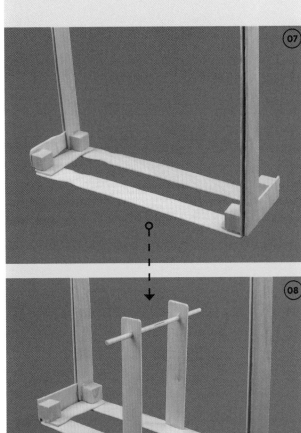

05 > Glue the four cubes to the corners of the base. Two of the cubes diagonally across from one another should be indented slightly from the edge to allow for the attachment of the bows in step 6.

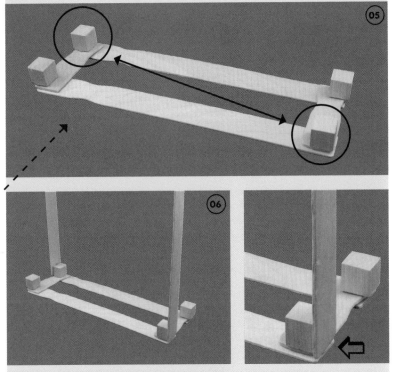

08 > Mark the center point on the long sides of the base and hot glue the two drilled 6" (15 cm) pieces of paint stirrer into place, directly across from one another. These will form the fulcrum. Cut a 5" (12.5 cm) piece of dowel. Insert the dowel through the drilled holes to make sure they're aligned.

06 > Glue a whole paint stirrer vertically to the two indented cubes. The flat edge of the bow should be aligned with the edge of the adjacent cube.

09 > Hot glue the split pieces of paint stirrer into place as trusses to strengthen the fulcrum supports.

10 > Line up three of the 3½" (9 cm) pieces of paint stirrer. Hot glue two more 3½" (9 cm) pieces on top to form a pad.

11 > Hot glue a whole paint stirrer to the center of the pad.

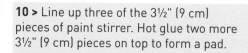
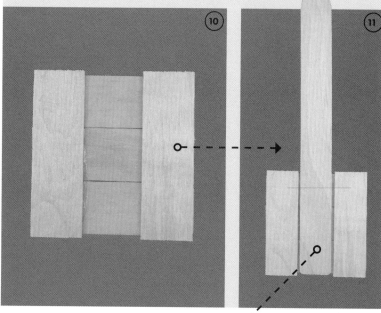

12 > Turn the pad over and use plenty of glue to attach the fulcrum dowel. The dowel should be centered on the pad.

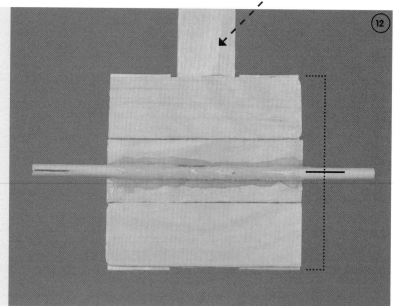

13 > Turn the pad dowel-side down. Carefully pry the fulcrum supports aside and insert the ends of the dowel through the drilled holes. The pad should fit snugly between the fulcrum supports, but there should not be any excessive friction that might hinder the catapult's performance.

14 > Cut a 45" (1 m 14.5 cm) piece of string. Fold it in half and tie a hitch knot around one side of the pad. Make sure that the string is hitched at the edge of the pad nearest the catapult arm. Attach the string to the pad with hot glue.

15 > Slightly bend the bow that corresponds to the string placement and fold the string over the top.

16 > Cover the string with hot glue while you hold it in place. When the glue has dried, wrap the glued string in place with duct tape to further secure it.

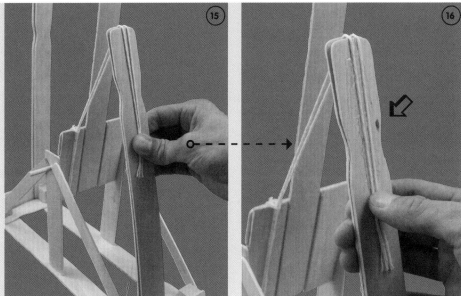

17 > Repeat steps 14 through 16 with an-other piece of string on the other side of the pad. This time, make sure the string is hitched at the bottom edge of the pad.

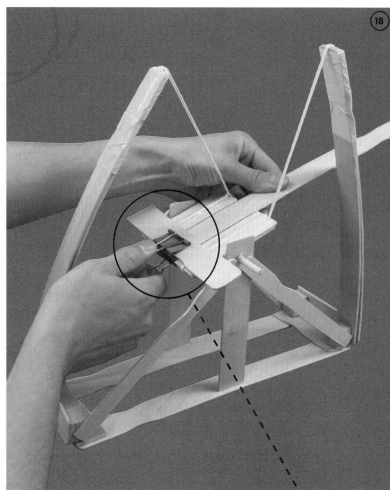

18 > Install the catapult's stop bar by bending the arm back and using the binder clip to attach the remaining 6" (15 cm) piece of paint stirrer.

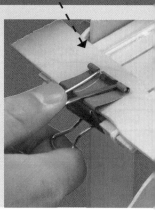

19 > Use glue and tape to attach a cup to the end of the catapult arm. Leave about ½" (1.3 cm) of paint stirrer exposed at the end.

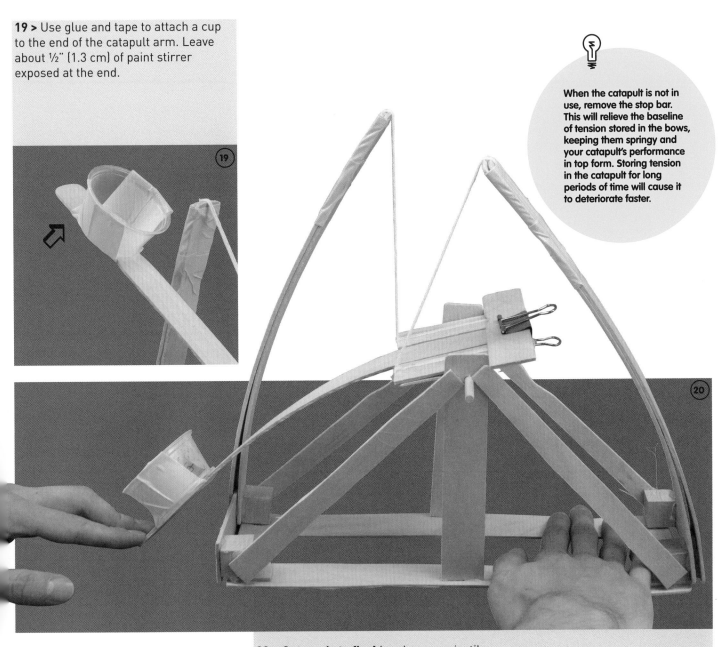

When the catapult is not in use, remove the stop bar. This will relieve the baseline of tension stored in the bows, keeping them springy and your catapult's performance in top form. Storing tension in the catapult for long periods of time will cause it to deteriorate faster.

20 > Get ready to fire! Load your projectile. Hold down the base of the catapult with one hand. Pull down on the very end of the catapult arm with your other hand and— **release!**

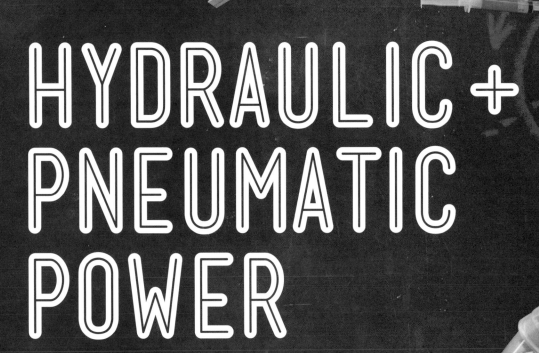

HYDRAULIC + PNEUMATIC POWER

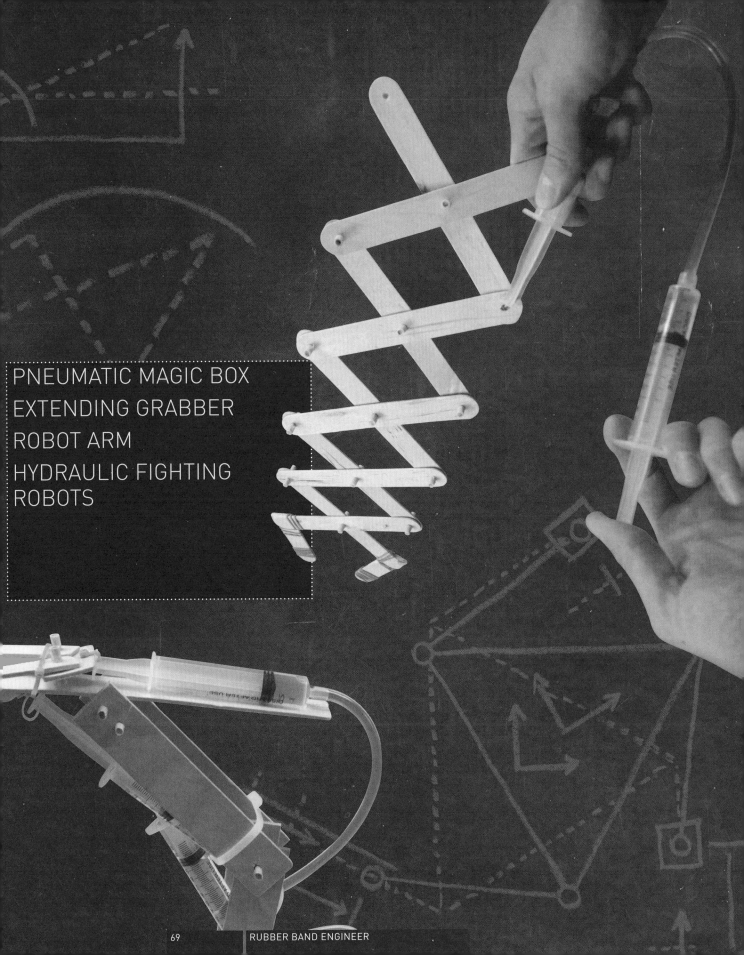

PNEUMATIC MAGIC BOX

EXTENDING GRABBER

ROBOT ARM

HYDRAULIC FIGHTING ROBOTS

HYDRAULIC AND PNEUMATIC SYSTEMS

THE BASICS

"Whoa, how does that work?!" is the question I always get when I show someone a pneumatic or hydraulic project. Here is the answer: Two plastic syringes filled with **water [*hydraulic*]** or **air [*pneumatic*]** and connected with a piece of tubing. When the full syringe's plunger is pressed, it forces the water or air into the other syringe, extending its plunger. The effect is surprisingly fluid [pun intended] and mechanically efficient.

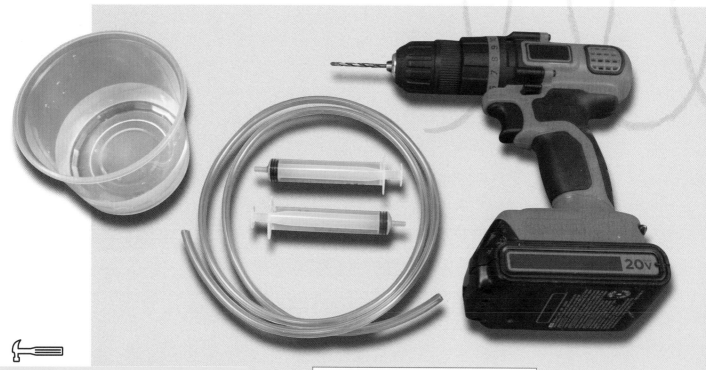

HYDRAULIC SYSTEM

All of the projects in this chapter use a simple pneumatic or hydraulic system, so first things first: learn how to build them.

TOOLS AND MATERIALS

TWO 10 ML LUER SLIP PLASTIC SYRINGES

VINYL TUBING WITH ⅛" (3 MM) INNER DIAMETER

CONTAINER OF WATER

DRILL WITH ⅛" (3 MM) BIT

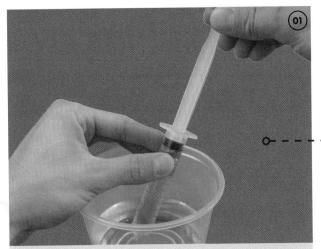

01 > Fill one syringe with water.

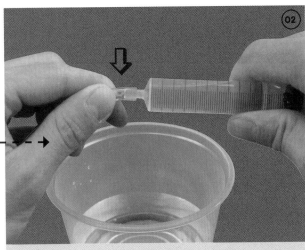

02 > Attach one end of the tubing to the syringe's nozzle.

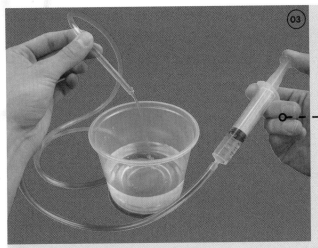

03 > Push water through the tubing until all of the air is removed.

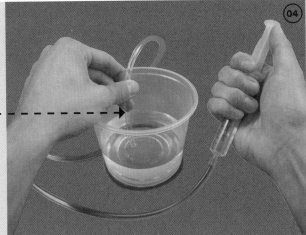

04 > Place the other end of the tubing into the water and fill the syringe up to 10 ml. Avoid overfilling the syringe.

MATERIAL SOURCING

Plastic syringes might seem like an unusual thing to find, but they're actually easy to come by. Ask your pharmacist if you can have some . . . for science. I use 10 ml syringes in this project, but other sizes could work if you scale the project accordingly.

Vinyl tubing can be found at aquarium supply stores and hardware stores.

HYDRAULIC + PNEUMATIC POWER

05 > Attach the loose end of the tubing to the other empty syringe making sure the plunger is all the way down.

06 > Try it out! When one syringe is pressed, the other will extend. Make sure there is little to no air in a hydraulic system.

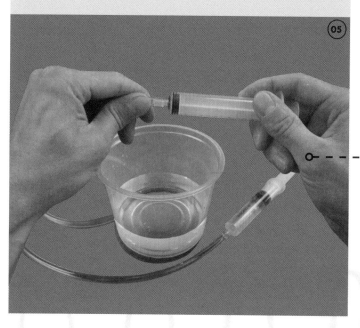

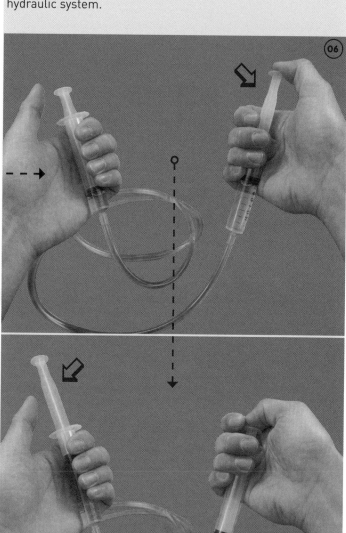

Some projects require a hole to be drilled on the end of the plunger; a cable tie can be threaded through the hole to attach the plunger to something. Use the ⅛" (3 mm) bit and drill into the corner of the plunger stem. Drill straight through the stem or up through the end of the plunger.

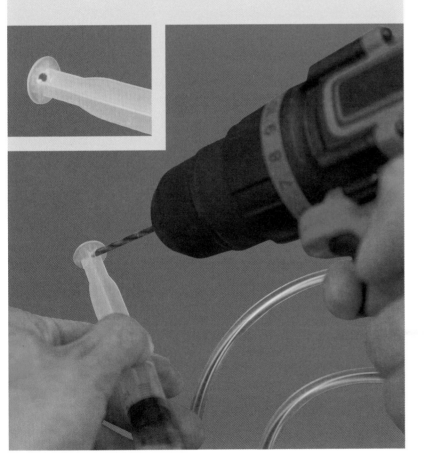

PNEUMATIC SYSTEM

The pneumatic system is basically the same idea: fill one syringe with air, leave one syringe empty, and connect the nozzles together with tubing. The pneumatic system is easier to make, however, it's less mechanically efficient. The air in the pneumatic system compresses or expands, whereas the water in a hydraulic system doesn't. The first project in this chapter has a very light load—a cardboard lid—and does not need the power of a hydraulic system.

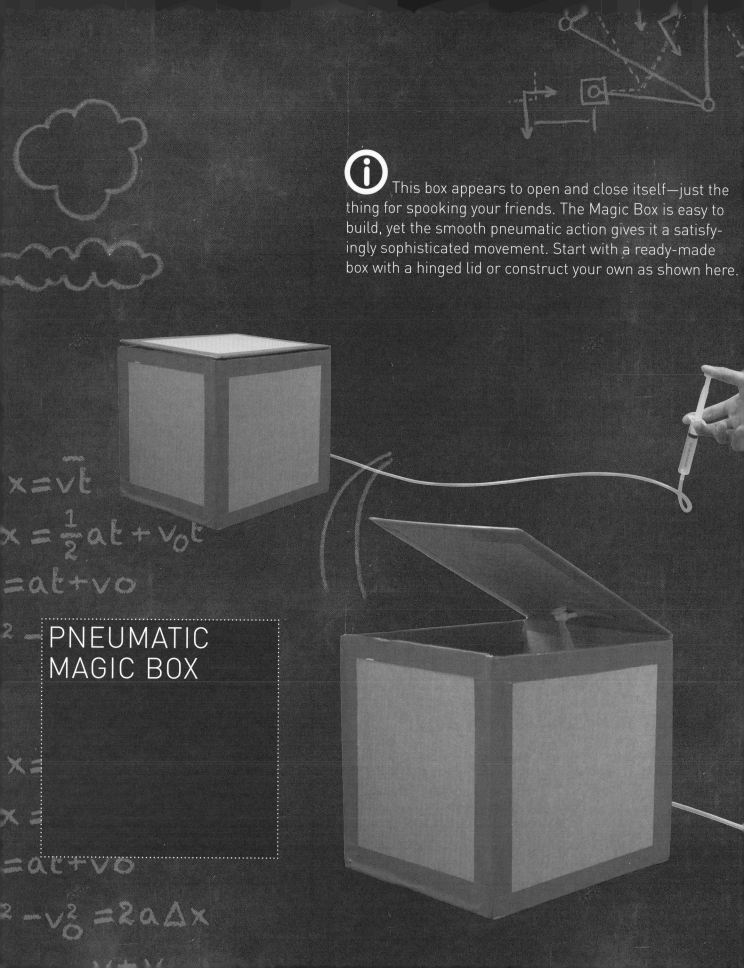

This box appears to open and close itself—just the thing for spooking your friends. The Magic Box is easy to build, yet the smooth pneumatic action gives it a satisfyingly sophisticated movement. Start with a ready-made box with a hinged lid or construct your own as shown here.

$x = \bar{v}t$

$x = \frac{1}{2}at + v_0t$

$= at + vo$

PNEUMATIC
MAGIC BOX

$= at + vo$

$^2 - v_o^2 = 2a\Delta x$

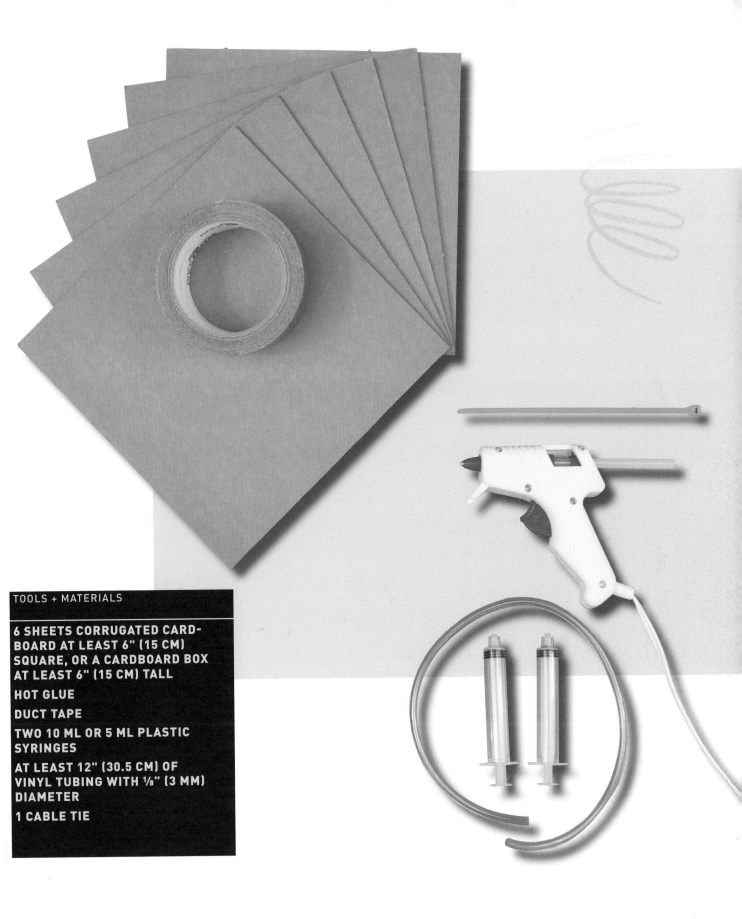

TOOLS + MATERIALS

6 SHEETS CORRUGATED CARD-BOARD AT LEAST 6" (15 CM) SQUARE, OR A CARDBOARD BOX AT LEAST 6" (15 CM) TALL

HOT GLUE

DUCT TAPE

TWO 10 ML OR 5 ML PLASTIC SYRINGES

AT LEAST 12" (30.5 CM) OF VINYL TUBING WITH ⅛" (3 MM) DIAMETER

1 CABLE TIE

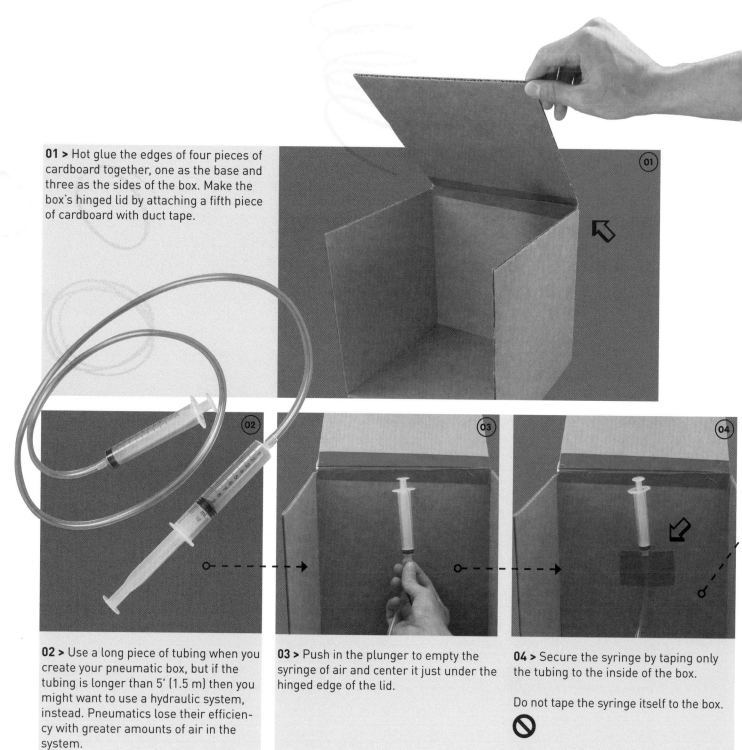

01 > Hot glue the edges of four pieces of cardboard together, one as the base and three as the sides of the box. Make the box's hinged lid by attaching a fifth piece of cardboard with duct tape.

02 > Use a long piece of tubing when you create your pneumatic box, but if the tubing is longer than 5' (1.5 m) then you might want to use a hydraulic system, instead. Pneumatics lose their efficiency with greater amounts of air in the system.

03 > Push in the plunger to empty the syringe of air and center it just under the hinged edge of the lid.

04 > Secure the syringe by taping only the tubing to the inside of the box.

Do not tape the syringe itself to the box.

05 > Wrap a cable tie around the syringe plunger.

06 > Extend the syringe's plunger. Lift the box's lid and tape the cable tie to the lid about 2" (5 cm) above the hinge. Cut off the excess cable tie.

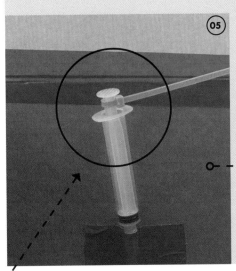

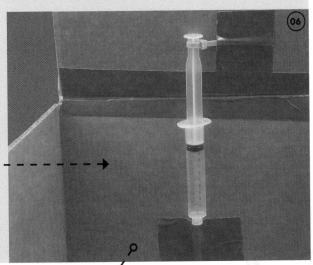

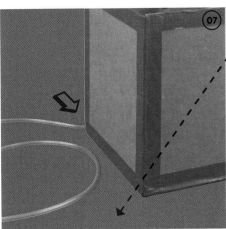

08 > Attach the second syringe and try it out! Once you have it working, invite an unsuspecting guest to inspect the box, and then magically open it before them. It's 100 percent sure to surprise!

07 > Make a discreet hole in a back corner of the box and lead the tubing through it. Glue the final piece of cardboard into place. The box is complete, and you may now cover up the corners and edges in duct tape.

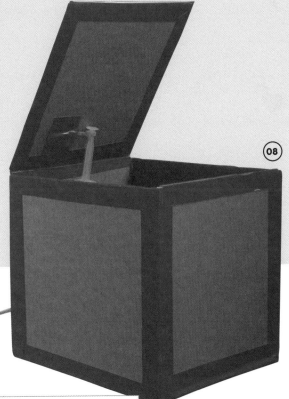

EXTENDING
GRABBER

ⓘ The crisscrossed extension mechanism is simple to build yet magical in effect. Customize how far it can reach and how to improve the grip. This design is best suited for poking office mates from a distance and lifting light loads. steps 1 through 8 offer you the hand-powered design. steps 9 through 11 tell you how to adapt the grabber to hydraulic power.

ELEVEN 6" (15 CM) CRAFT STICKS

MASKING TAPE

DRILL WITH ⅛" (3 MM) BIT

TWO ⅛" (3 MM)-WIDE BAMBOO SKEWERS

WIRE CUTTER

4 RUBBER BANDS

TWO 10 ML LUER SLIP SYRINGES (OPTIONAL)

2 CABLE TIES (OPTIONAL)

VINYL TUBING WITH ⅛" (3 MM) INNER DIAMETER (OPTIONAL)

DUCT TAPE

MATERIAL SUBSTITUTIONS

WIRE CUTTER	**Any tool that can cleanly cut skewers, such as kitchen scissors or a utility knife.**
6" (15 cm) CRAFT STICKS	**Paint stirrers, square dowels, or any rigid wood that can be drilled.**

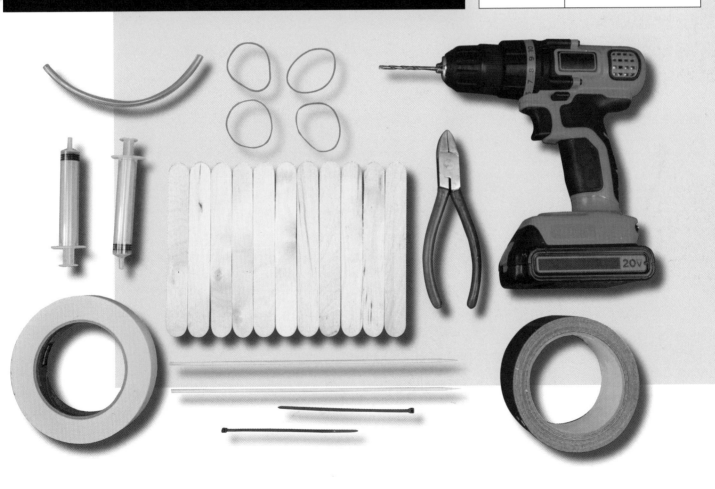

01 > Set one craft stick aside. Stack the other 10 craft sticks one on top of the other with all edges aligned and bundle them together with masking tape. Drill three holes through the bundle: one in the center and one at each end. Drill straight down with light pressure to avoid splitting the sticks. Remove the masking tape.

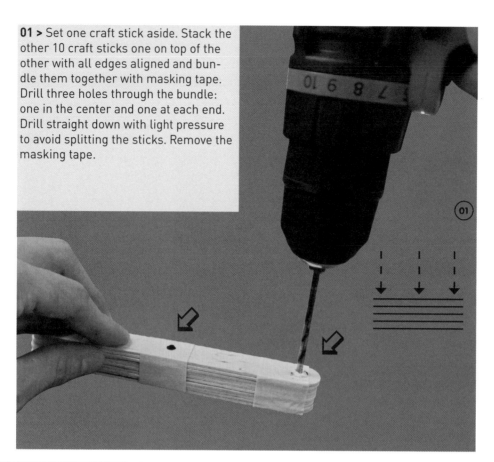

02 > Arrange the 10 sticks in the crisscross pattern called for in this project. **Note the consistent layering pattern.**

03 > Lift the first two sticks and push a skewer through the center hole.

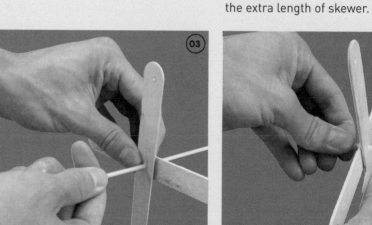

04 > Wrap masking tape around both ends of the skewer to prevent it from sliding out. Use wire cutters to cut off the extra length of skewer.

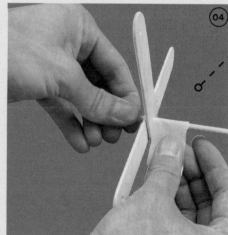

05 > Give the extending mechanism a test after joining the first four sticks. The mechanism should extend in a straight line. If the extender is leaning to one side, it means that the holes were drilled unevenly.

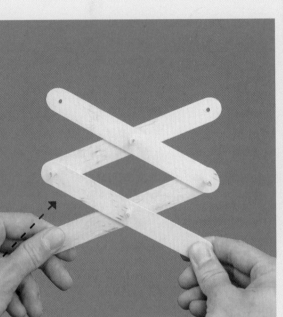

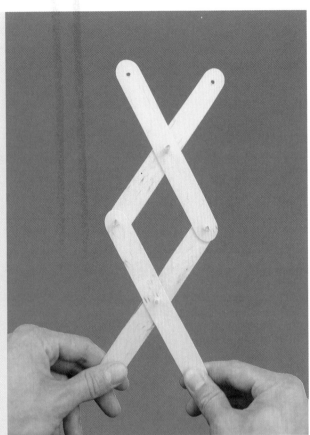

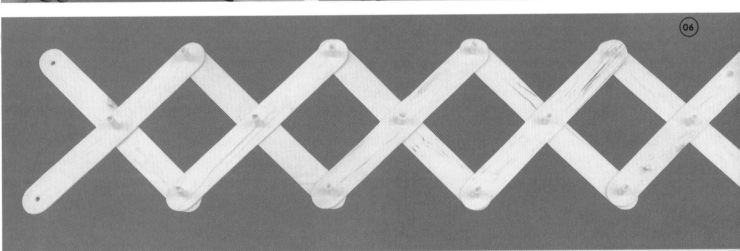

06 > Continue joining the sticks, trimming the ends of the skewers as you go. The more sticks you add, the longer the extender can reach. Note that the extender loses mechanical efficiency the longer it becomes.

07 > Make the grabber claw. Take the remaining craft stick set aside in step 1 and cut it in half across its width with a craft knife. Tape the two pieces into place at one end of the extender.

They should angle toward one another slightly when the arm is contracted, and should overlap slightly when the arm is extended.

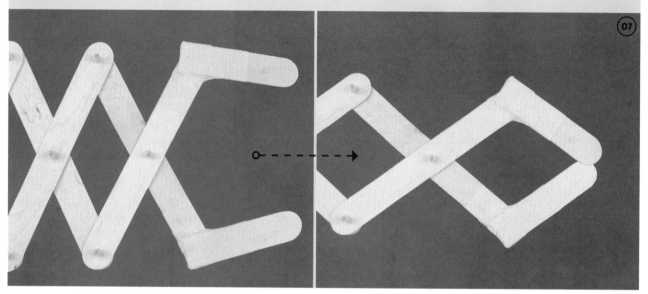

08 > Enhance the claw's grip by wrapping rubber bands around the tip and put it to work!

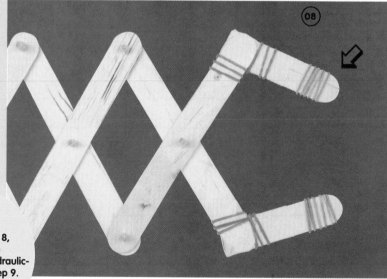

You can end the project with step 8, or you can make this extender hydraulic-powered with step 9.

09 > Attach the hydraulic system. Contract the extender fully, and remove one of the skewer pins as shown. Drill a hole into the plunger tip. Thread two cable ties through the holes in the craft sticks to attach the syringe.

11 > Test it out! It should be able to fully extend and contract most of the way. You can further calibrate the mechanism by adjusting the position of the output syringe. Tape the syringes together for easier control.

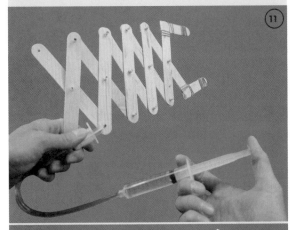

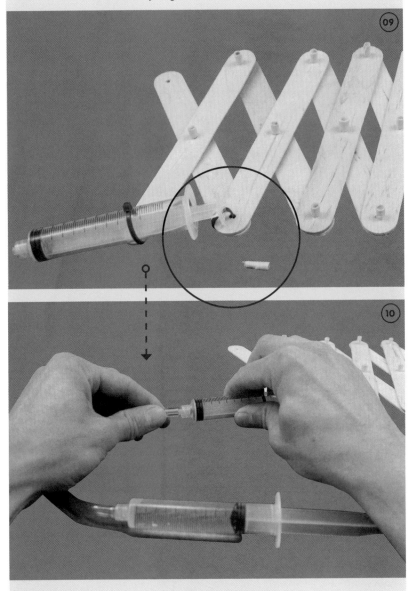

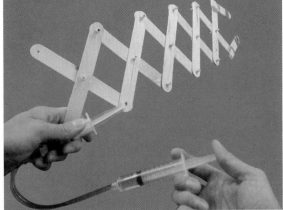

10 > Attach the tubing and a water-filled syringe to complete the hydraulic system. Follow the steps for hydraulic hook-up on page 70.

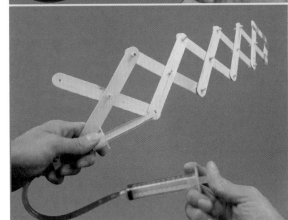

Operate a robotic arm by masterfully manipulating an array of hydraulic controls. This design has no less than five hydraulic inputs, giving you a high degree of precision and articulation. This is one of the most challenging projects this book has to offer, but when up and running, it's one of the most impressive to witness.

ROBOT ARM

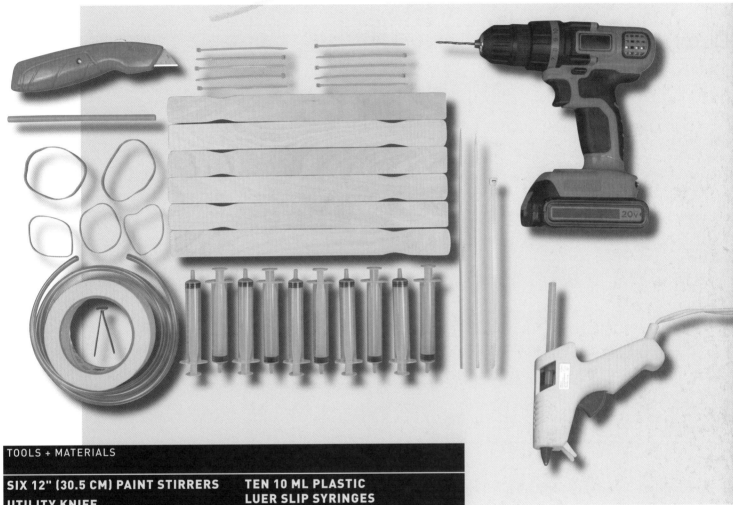

SIX 12" (30.5 CM) PAINT STIRRERS

UTILITY KNIFE

MEASURING TAPE

DRILL WITH ⅛" (3 MM) BIT

HOT GLUE GUN

TWO ⅛" (3 MM)-WIDE BAMBOO SKEWER

MASKING TAPE

PLASTIC DRINKING STRAW

TEN 10 ML PLASTIC LUER SLIP SYRINGES

4' (1 M 22 CM) VINYL TUBING WITH ⅛" (3 MM) INNER DIAMETER

10 CABLE TIES

POSTER BOARD

BRASS PAPER FASTENER

3 RUBBER BANDS

2 THICK RUBBER BANDS

01 > Prepare the articulating arm pieces.

Use the utility knife to cut the following from four paint stirrers:

Seven 6" (15 cm) pieces

Two 3" (7.5 cm) pieces

The holes for pieces 2, 4, 5, and 6 are drilled about ½" (1.3 cm) from the ends.

The number 6 pieces have an additional hole drilled ½" (1.3 cm) from the first at one end.

Pieces 1, 3, and 5 have a hole drilled in the center.

NOTE: As you prepare the pieces, lightly number them with a pencil.

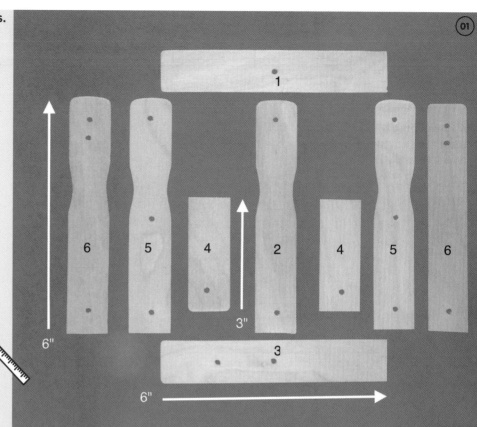

02 > Prepare the pieces for the claw.

Use the utility knife to cut the following from the remaining two paint stirrers:

One 8" (20.5 cm) piece

One 4" (10 cm) piece

Two 3" (7.5 cm) pieces

Two 2" (5 cm) pieces

One 1.5" (4 cm) piece

03 > Split the 1.5" (4 cm) piece, two 3" (7.5 cm) pieces, and the 4" (10 cm) piece in half lengthwise. Pieces 8, 11, and 12 have holes drilled ¼" (6 mm) from the ends. The number 9 pieces have a hole drilled ¼" (6 mm) from one end and a hole drilled in the center.

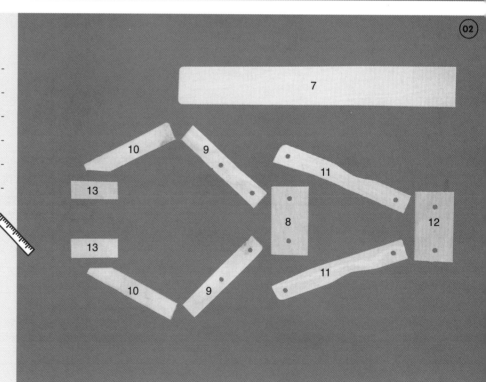

04 > Assemble the articulating arm. Hot glue an H-shaped base using pieces 1, 2, and 3 from step 1, as shown.

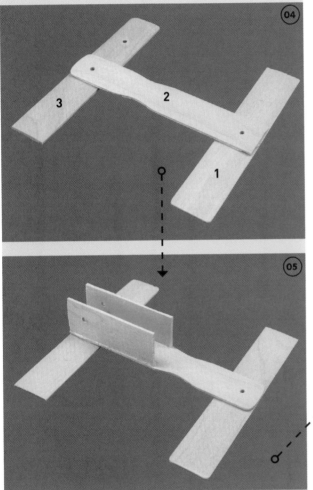

05 > Hot glue the two number 4 pieces from step 1, onto the base, as shown.

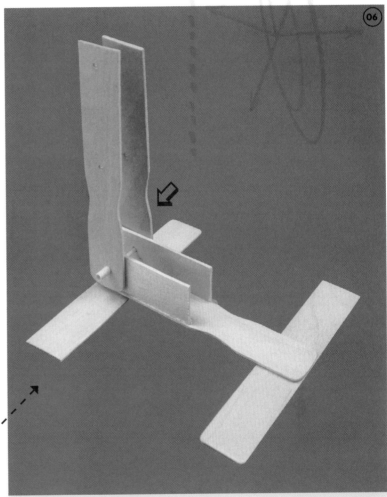

06 > Select the two number 5 pieces from step 1 and connect them to the base with a skewer. This will act as a hinge. Wind tape around the ends of the skewer to keep it in place. Trim off the extra length of skewer.

07 > Take the two number 6 pieces from step 1 and line up their holes with those at the top of the hinge. This time, before threading a skewer through the holes, cut a piece of plastic straw the width of the distance between the holes.

The straw ensures that the arm pieces are spaced correctly. Do the same at the opposite end. Tape the ends of the skewers and trim the excess.

08 > Assemble the claw. Select pieces 7 and 8 from step 2. Hot glue them together as shown, making sure the drilled holes are not blocked.

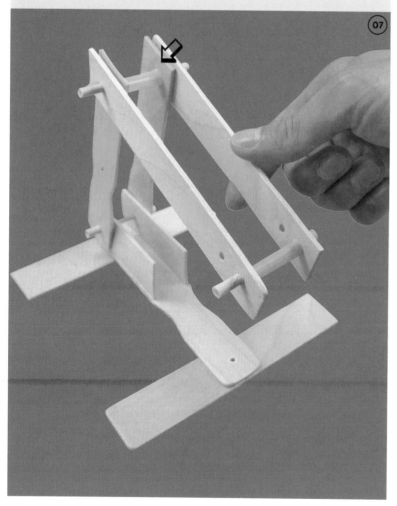

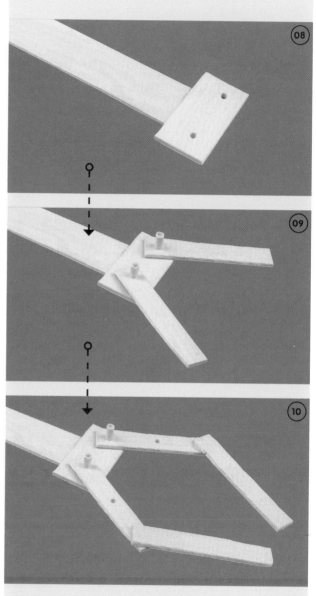

09 > Attach the two number 9 pieces to the claw base with skewer hinges.

10 > Hot glue the two number 10 pieces into place at an obtuse angle, as shown.

11 > Attach the two number 11 pieces from step 2 to either side of the number 12 piece with skewer hinges. Place this device on top of the prepared section of claw and attach the number 11 pieces to the number 9 pieces with skewer hinges as shown. When complete, test for articulation. The tips of the claw should come together without any interference.

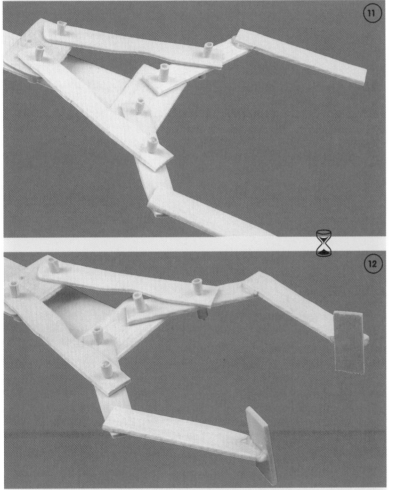

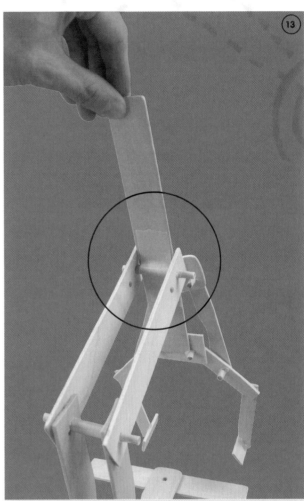

13 > Hot glue and tape the claw assembly onto the front straw axle of the articulating arm, as shown.

12 > Hot glue the two number 13 pieces securely onto the tips of the claw. In this example, the tip has been cut at an angle to allow the tips to press together when the claw closes.

14 > Now you will need to create five hydraulic systems using ten syringes. The length of the tubing that connects them will depend on how far away you want the controls to be, though you'll need at least 1' (30.5 cm) for each. (See page 70 for hydraulic system details.)

16 > Cable tie the first syringe to the base and to the newly inserted skewer.

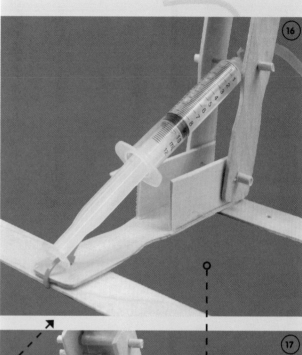

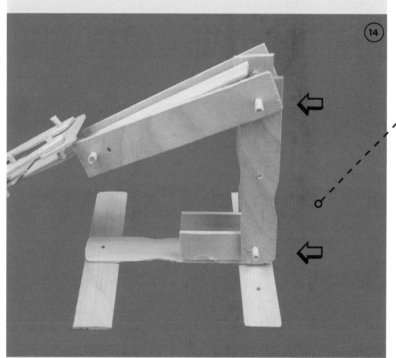

15 > Insert a skewer through the central holes of the two upright sections of the articulating arm. Tape and trim the skewer ends.

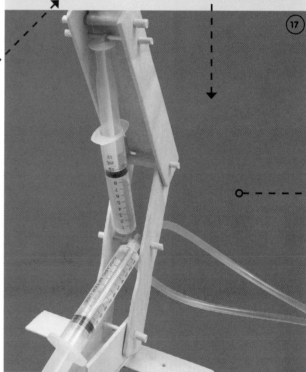

17 > Attach the nozzle of the second syringe to the same skewer. Insert another skewer through the holes next to the straw hinge as shown, taping and trimming the ends. Connect the plunger of the second syringe to it with a cable tie.

19 > Attach the fourth syringe to the cable tie wrap and the arm hinge as shown.

18 > Open the claw wide. Position the third syringe on top as shown and attach it with a cable tie. Glue the syringe casing into place. Wrap another cable tie around the claw to prevent it from bulging upward.

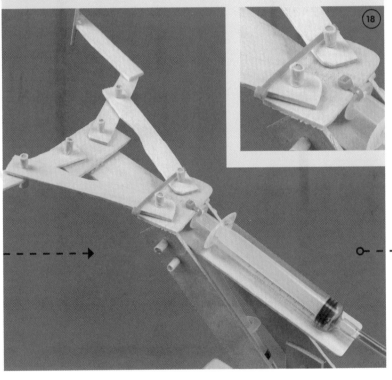

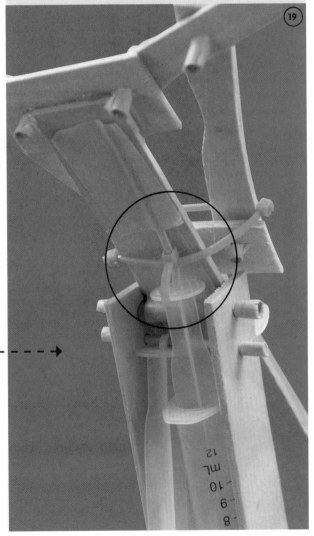

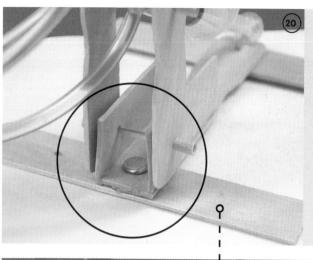

20 > Create a working surface for the articulating arm. Center and punch a hole in the poster board about 6" (15 cm) from one end. Insert the brass fastener through the hole in the articulating arm's base and the hole in the poster board. Tape the fastener open on the underside of the board.

PRACTICE AND ADJUSTMENTS

Remain determined if your robot arm doesn't work perfectly at first. It takes some time to get the hang of operating the controls, and you may need to make small adjustments to get the best performance.

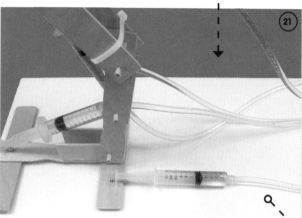

21 > Finish installing the hydraulic outputs by attaching the fifth syringe to the remaining hole in the articulating arm's base with a cable tie. This syringe will push and pull on the base, causing the robot arm to turn from side to side.

22 > To make sure that your robot turns left and right evenly, point the base of the robot arm straight forward, then fill the syringe halfway. Hold the output in place with your hand and test it. The arm should turn left and right an equal amount. When satisfied, punch a hole on either side of the nozzle and tie it to the poster board. Cover the connection with hot glue to prevent the holes in the poster board from tearing over time.

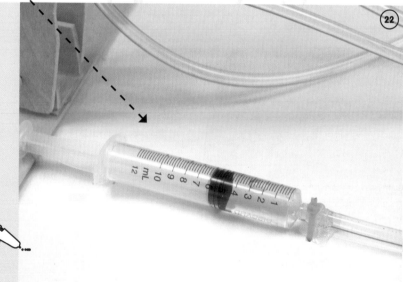

24 > Manipulate the controls to pick up and move objects with precision.

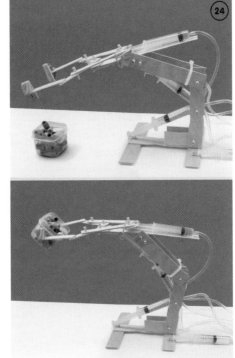

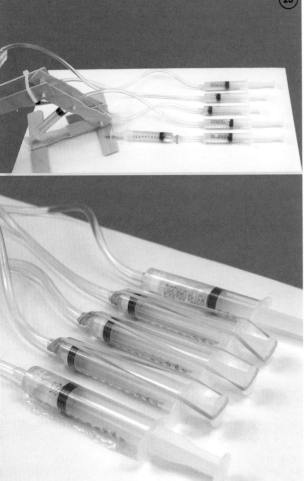

23 > Arrange the inputs on the remaining area on the poster board. Loop rubber bands around the controls that raise and lower the robot arm. Without the rubber bands, the weight of the arm will push against the outputs and cause the whole arm to drop. **The robot arm is done!**

NOW GET TINKERING

The tip of the arm can be designed in many different ways for a variety of applications. The entire grabbing apparatus can be modified to pick up specific objects or simply improve the grip. In this example, two thick rubber bands have been wrapped around the tips to give the grabber better grip.

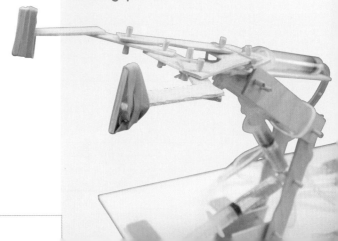

HYDRAULIC
FIGHTING
ROBOTS

ⓘ Yes, it's just what it sounds like. Here we will design, build, and customize our own hydraulic-powered battling robots! Each robot begins with a basic foundation, but after that it's up to you to decide how to build to win. Flip opponents over, push them off the table, or devise another wickedly clever way to dominate the battlefield.

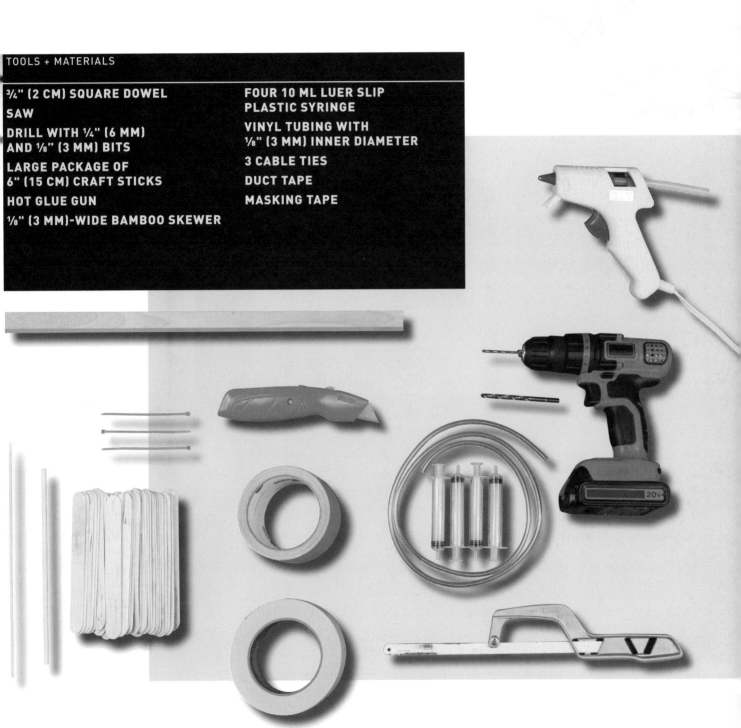

¾" (2 CM) SQUARE DOWEL

SAW

DRILL WITH ¼" (6 MM)
AND ⅛" (3 MM) BITS

LARGE PACKAGE OF
6" (15 CM) CRAFT STICKS

HOT GLUE GUN

⅛" (3 MM)-WIDE BAMBOO SKEWER

FOUR 10 ML LUER SLIP
PLASTIC SYRINGE

VINYL TUBING WITH
⅛" (3 MM) INNER DIAMETER

3 CABLE TIES

DUCT TAPE

MASKING TAPE

01 > Robot Foundation

Use the saw to cut cubes from the ¾" (2 cm) dowel. You'll need at least four per robot. Drill a ¼" (6 mm) hole through the center of four cubes. For stability, use a vice or a pair of pliers to grip the cube while you drill.

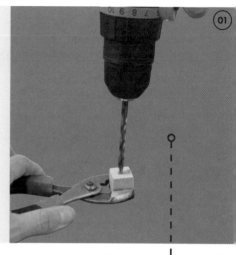

MATERIAL SOURCES

Some craft stores sell packs of cubes with holes, often containing short wooden rods. You can also buy cubes with holes directly from an online vendor listed in Material Sources at the back of the book.

02 > Begin the base by gluing three craft sticks into an H shape. Glue one of the cubes with a hole in the center, and then glue an 8" (20.5 cm) piece of ¼" (6 mm) dowel into the hole.

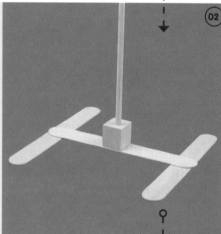

03 > Slide two more of the drilled cubes over the dowel to begin forming the pivot column. Hot glue a craft stick to each side of the pivot cubes—position one pivot cube at the top of the craft sticks and one at the bottom. Make sure that the craft sticks do not overlap the fixed cube at the base of the column.

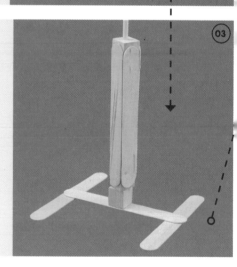

04 > Create a hydraulic attachment appendage at the bottom of the pivot column. Use hot glue to attach two 3" (7.5 cm) pieces of craft stick and the final drilled cube.

05 > Wrap tape around the top of the dowel and trim it, leaving about ½" (1.3 cm) at the top. This will prevent the pivot column from slipping off the dowel during combat.

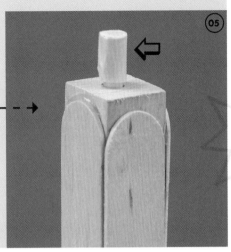

06 > Stack 4 craft sticks on top of one another and tape them together. Drill a hole through one end of the stack and remove the tape.

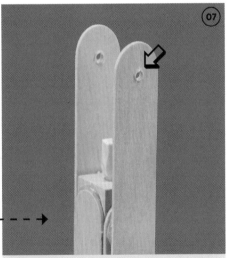

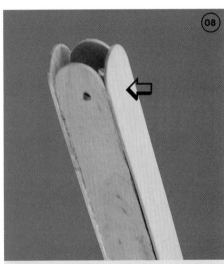

07 > Hot glue two of the drilled craft sticks to the top of the pivot column to create a hinge. Glue the sticks to the same sides of the column as the arms for the hydraulic attachment appendage from step 5. Position the holes in the craft sticks about 2" (5 cm) inches above the top of the column. Insert a skewer into the holes as you build to make sure that the holes are aligned.

08 > Begin building the fighting arm by creating a rectangle box using the remaining drilled craft sticks from step 7 and two undrilled craft sticks. Hot glue the edges of the four craft sticks at right angles. Make sure that the drilled holes are directly opposite one another.

09 > Align the holes in the fighting arm with the holes at the top of the pivot column. Slide a skewer through the four holes. Wrap tape around the ends of the skewer and trim off the excess.

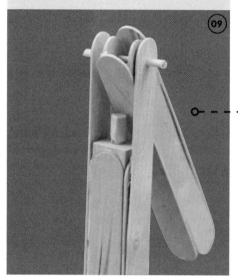

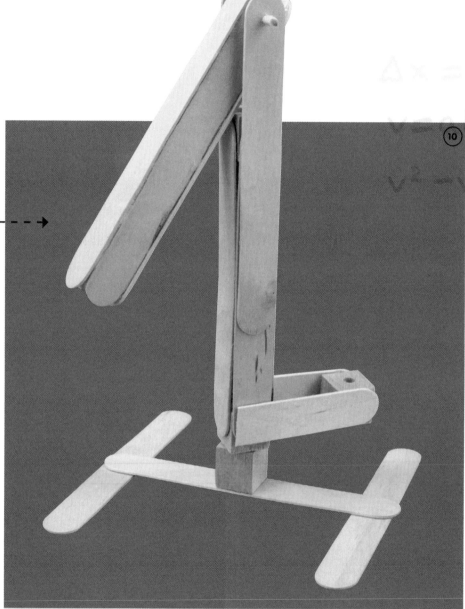

10 > The basic robot foundation is complete! From here on, it's up to you to invent a design that will dominate the battlefield. The following images are just examples of how the base and the fighting arm can be modified to win.

EXAMPLE DESIGN 1

A_Strategically positioned globs of dried hot glue on the underside of the base to give the robot better traction.

B_The fixed base cube is reinforced to prevent it from leaning or breaking off in combat.

C_The base is widened for added stability. This fighting robot is designed to lean forward and push opponents off the table, so the pivot column (and thus the center of mass) is intentionally placed near the front of the base.

D_The arm is extended to better reach the opponent's pivot column.

E_A wide and sturdy fork made of glued craft sticks allows you to push opponents as opposed to flipping them.

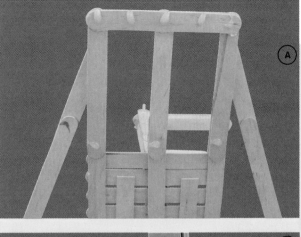

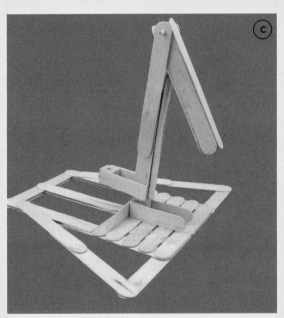

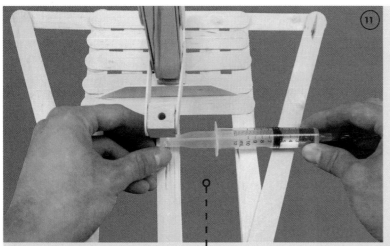

11 > Attach the hydraulics.
To attach the pivoting hydraulic system, center the pivot column. Position the output syringe so that it is halfway extended.

12 > Secure the syringe in place by tightly wrapping it around the base with duct tape.

13 > Thread a cable tie through the hole in the cube and the hole drilled in the syringe plunger.

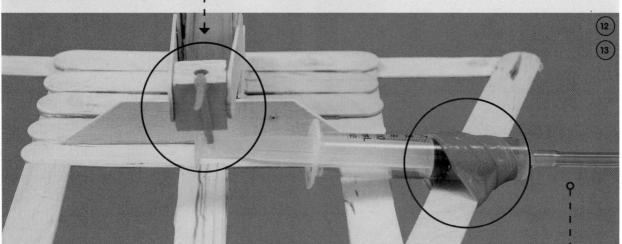

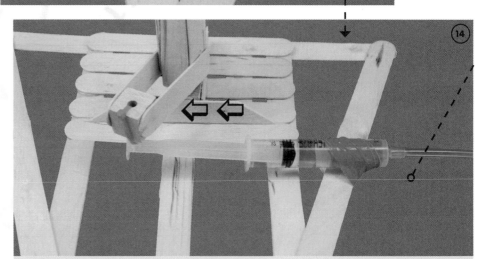

14 > Test the calibration by fully extending and retracting the output. The robot should turn equally in both directions.

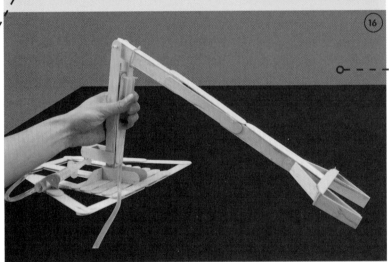

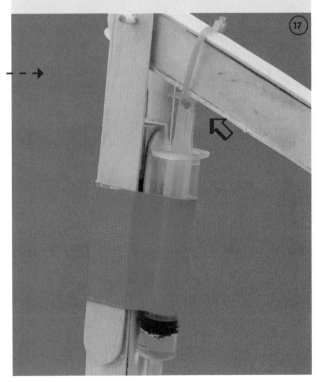

17 > Tightly wrap duct tape around the syringe and the column. The fighting arm should rest on the table while the syringe has 2 to 3 ml of water in it. This gives you the option to apply some downward force even while the arm is touching the tabletop. This can be advantageous in some situations, especially for robots that are designed to flip opponents.

15 > To attach the fighting arm hydraulic system: Tie the output with a cable tie to the underside of the arm near the column. It should be tied a little loosely. (You may need to link two 4" (10 cm) cable ties to make it long enough.)

16 > Hold the output against the column while the arm is resting on the table. Test it by extending it fully. Note the position of the syringe relative to the column.

18 > Finish the robot by tidying up the
tubing and taping the controls together.
Check to make sure that the tubing is
long enough to allow the robot plenty
of mobility.

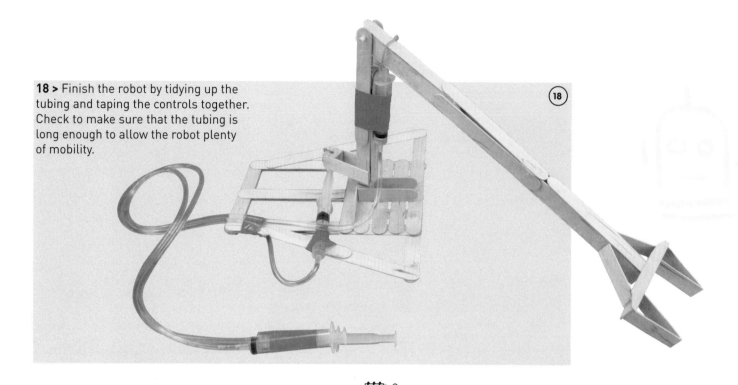

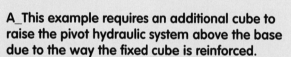

EXAMPLE DESIGN 2

A_This example requires an additional cube to
raise the pivot hydraulic system above the base
due to the way the fixed cube is reinforced.

B_In this example, the weapon is designed to
be wedged under the opponent's base to flip
it over!

C_The pivot column is positioned farther back
on the base. This design does not lean forward,
allowing it to remain stable when it flips the
opponent.

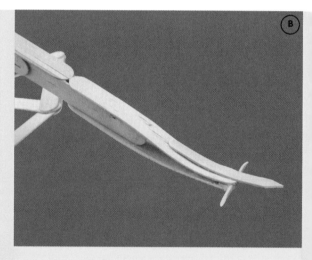

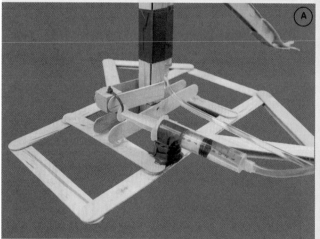

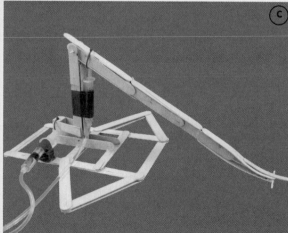

Suggested Battle Rules

A_Place both robots on a table with a gap of at least 3" (7.5 cm) between them. Start with both robots facing forward and fighting arms raised to the maximum height.

B_Countdown from 3, 2, 1—fight! Manipulate the controls and try to flip the opponent over, or push it off the table. Avoid trying to control the robot by pulling on the tubing.

C_The orange robot has succeeded in pushing its opponent near the edge of the table.

D_Now for the final blow!

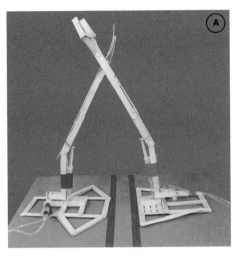

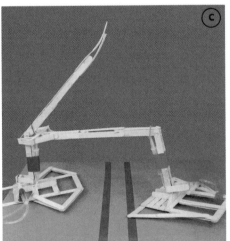

CREATIVE CONSTRAINTS

As with any competitive game, players are given constraints to prevent exploitation. Follow these guidelines for richer designs and skill-based battles:

Limit building materials to 6 cubes, 75 craft sticks, and 1 skewer. This includes materials that are required for the foundation. This encourages ingenuity versus simply piling 500 sticks onto the back of your robot to add weight.

The base must fit within a 12" (30.5 cm) square. The arm may extend outside of that area. This prevents sprawling bases that are impossible to defeat.

Incorporate your own material choices for a unique experience, but limit how much can be used.

REDESIGN YOUR ROBOT

The battle doesn't end here. Take your bot back to the workbench and redesign it to be an even more effective fighter, and then fight again!

FLIGHT

IMPROVISED DARTS
SLINGSHOT ROCKET
RUBBER BAND
HELICOPTER

(i) Darts may be the simplest project in this book, but don't be fooled—creating a super-effective dart is more nuanced than you might think. This project isn't a step-by-step plan, but rather a guide to illustrate what factors contribute to—and detract from—a great improvised dart, whether made from a pen, a pencil, or a drinking straw.

IMPROVISED
DARTS

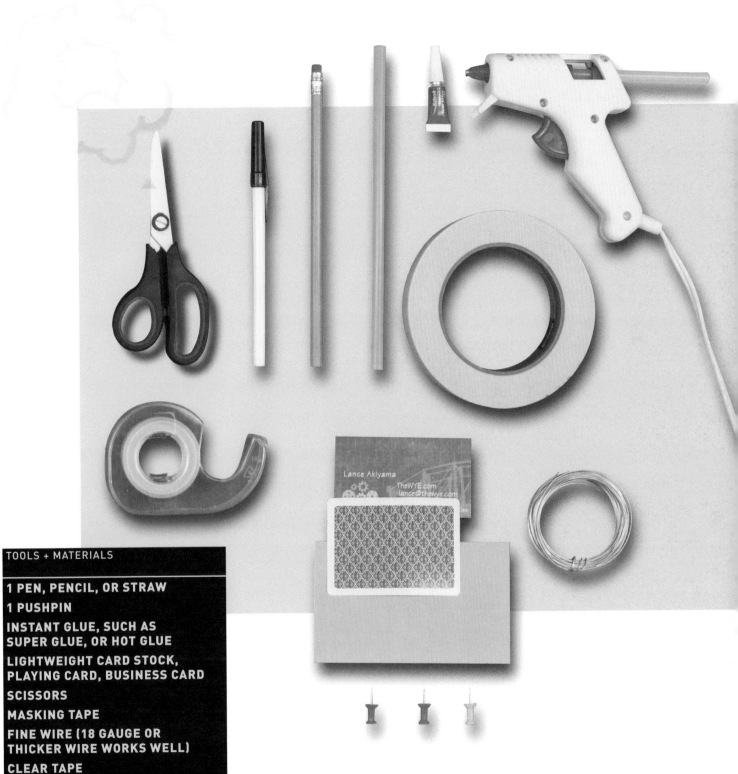

TOOLS + MATERIALS

1 PEN, PENCIL, OR STRAW

1 PUSHPIN

INSTANT GLUE, SUCH AS SUPER GLUE, OR HOT GLUE

LIGHTWEIGHT CARD STOCK, PLAYING CARD, BUSINESS CARD

SCISSORS

MASKING TAPE

FINE WIRE (18 GAUGE OR THICKER WIRE WORKS WELL)

CLEAR TAPE

Pen Darts

A pen shaft is suitable for dart making because the casing is fairly light and it's easy to install a pushpin in the end.

01 > Disassemble the pen. The only piece you'll need is the casing. The other components will add too much weight to the dart.

02 > Glue the pushpin inside the pen casing. If the pen casing is irregularly shaped, insert the pushpin into the heavier end.

03 > Cut 2 or 3 small triangular fins from the card stock and tape them to the back end of the dart. Tape each fin on both sides to make sure it stands perpendicular to the dart shaft. Fins are necessary to stabilize the dart. The straighter the fins, the straighter the dart will fly. Use materials that are light yet rigid. If you use card stock or cardboard that is too heavy, it will pull the dart off balance. It is best to have most of the weight toward the front.

04 > Wrap several rounds of wire around the front of the dart. This creates a leading weight, which will help keep the dart pointed forward. (See page 34 for more information about this crucial factor.) If wire is not available, fill the tip of the pen casing with hot glue.

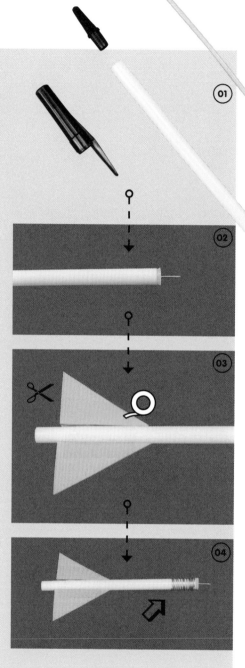

Pencil Darts

Pencils are not ideal dart material because they are dense and heavy, but they can be used in a pinch.

01 > Hot glue a pushpin to the eraser. It might seem counterintuitive to make the eraser the front of the dart, but this end is heavier than the other, which will contribute to the leading weight.

02 > Secure the pushpin further by wrapping it with masking tape.

03 > Wrap wire or another dense material around the front of the dart to enhance the leading weight.

04 > Create fins for the dart as in Step 3 of **Pen Darts.** This example uses fins made from a playing card. Make sure that the fins are straight.

Straw Darts

Lightweight plastic straws are perfect for improvised darts because the leading weight has a profound effect on its trajectory.

01 > Glue a pushpin into at one end of the straw.

02 > Further secure the pushpin with making tape.

03 > Create fins for the dart as in step 3 of **Pen Darts.** The fins shown here are made of a business card, which is very thin but also very rigid.

04 > Wrap the front end of the dart with wire to add leading weight. Secure the ends of the wire with glue.

05 > Set up a corrugated cardboard target and see what you can do. Adjust the leading weights of your darts as necessary for better control.

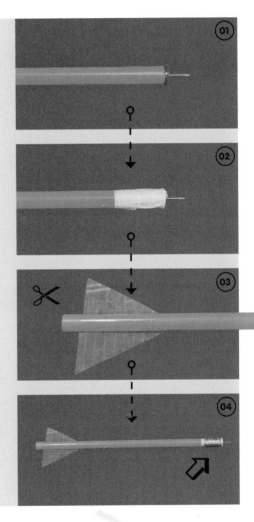

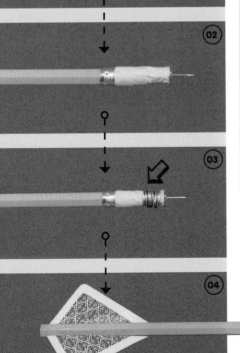

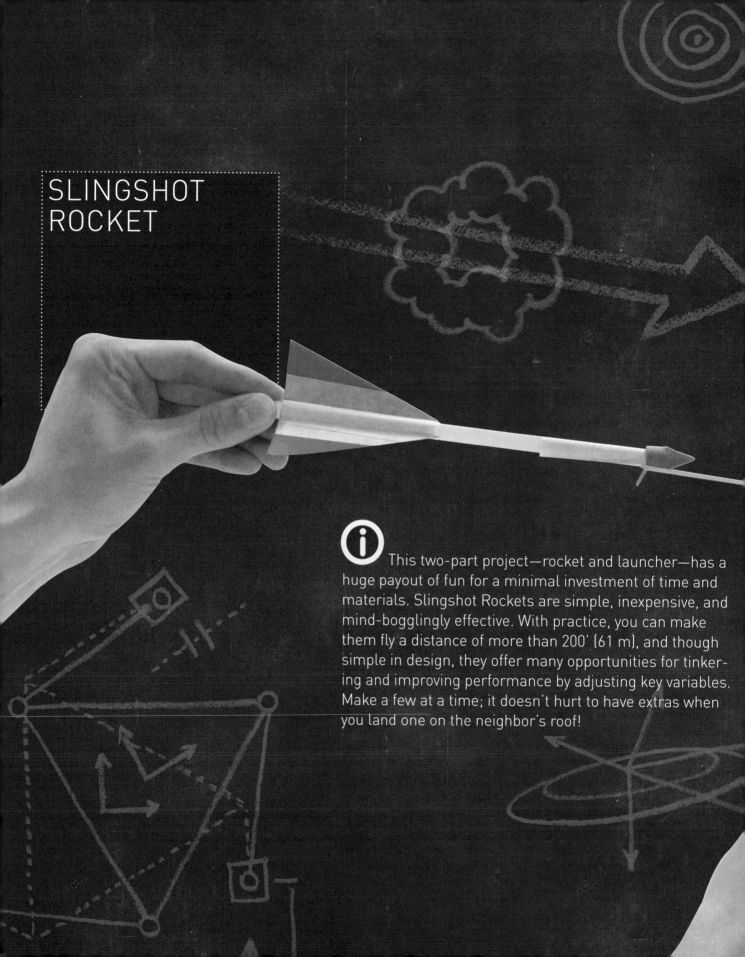

SLINGSHOT ROCKET

This two-part project—rocket and launcher—has a huge payout of fun for a minimal investment of time and materials. Slingshot Rockets are simple, inexpensive, and mind-bogglingly effective. With practice, you can make them fly a distance of more than 200' (61 m), and though simple in design, they offer many opportunities for tinkering and improving performance by adjusting key variables. Make a few at a time; it doesn't hurt to have extras when you land one on the neighbor's roof!

1 PENCIL-CAP ERASER
1 LARGE STRAW
1 PAPER CLIP
PLIERS
1 INDEX CARD
1 PENCIL
2 RUBBER BANDS
SCISSORS
MASKING TAPE

MATERIAL SUBSTITUTIONS

LARGE STRAW	**Ballpoint pen tube or 3 thin straws bundled and taped together**
PENCIL	**Craft stick, ballpoint pen, or another sturdy object at least 4" (10 cm) long**
PENCIL-CAP ERASER	**1" (2.5 cm) piece of a mini hot glue stick; 1" (2.5 cm) piece cut from the eraser end of a pencil; or other small, dense, rubbery material**
INDEX CARD	**Card stock; thin cardboard. Avoid using thin printer paper.**

01 > Make the Rocket

Fit the eraser onto the end of the straw to weight the rocket's tip. This additional weight will help propel the rocket to greater distance and help cushion a crash landing.

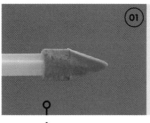

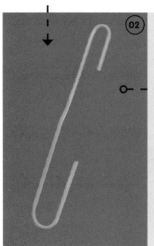

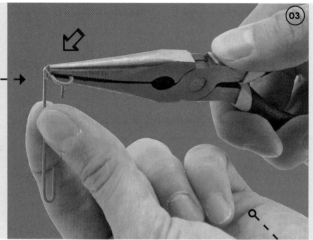

02 > Bend the paper clip into an elongated C shape.

03 > Use the pliers to bend one curved end of the paper clip into a 30-degree angle. This will be the hook for the rubber band.

04 > Position the paper clip on the straw so that the hook stands up directly below the eraser. Use masking tape to attach the remaining length of the paper clip to the straw. Wrap the paper clip in place tightly so it won't come loose when the rocket is launched!

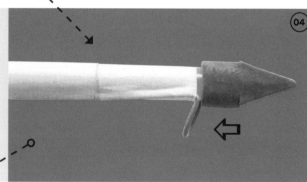

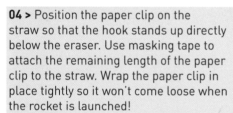

05 > Measure and cut 3 or 4 fins from the index card. Cut the fins as right angles, approximately 2" × 1½" × 1" (5 × 4 × 2.5 cm). The fins will help stabilize the rocket's flight path.

DESIGN VARIABLES: FINS

The fin size will alter the rocket's performance and maximum distance. Smaller fins will create less drag and allow the rocket to go farther, but if the fins are too small they will not stabilize the rocket's flight path. Larger fins produce more drag, but the rocket will be very stable. Very large fins create a rocket that glides gracefully. No matter the size, the fins must be aligned with the shaft to be effective.

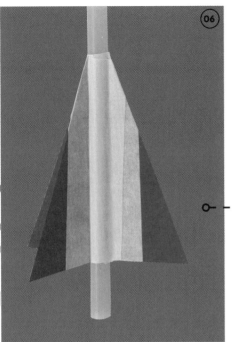

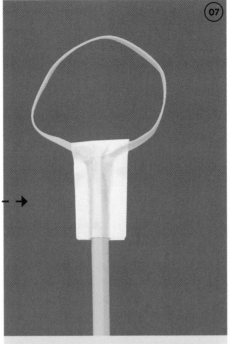

06 > Line up one fin so that the shortest side is at a right angle to the straw. Use a piece of tape that is as long as the fin to attach it to the straw. Check to make sure the fin is straight. Cut away or fold down any excess tape. Use a second piece on the other side of the fin to ensure that it doesn't wobble too much while in flight. Repeat with the other fins, spacing them evenly. Leave about ½" to 1" (1.3 cm to 2.5 cm) of exposed straw behind the fins so that you will have space to grip the rocket when you launch it.

07 > Make the Slingshot Launcher
Use one or two layers of masking tape to attach one end of the rubber band to one end of the pencil. The rubber band must be secured to the very end of the pencil to prevent the rocket from colliding with the slingshot. Now you're ready for blast off!

HOW TO LAUNCH

Hold the slingshot perpendicular to the ground with your thumb against the pencil.

Hold the rocket with your other hand, pinching the end of the rocket between the fins.

Latch the rocket's paper clip hook onto the rubber band.

Hold the slingshot out in front of you. Draw the rocket back toward yourself to build up the full potential of elastic energy. The rubber band should be perpendicular to the slingshot.

Allow the slingshot to tilt forward as you release the rocket. This will help prevent the rocket from colliding with the slingshot.

Add an extra boost to the launch by flicking the slingshot forward as the rocket is launched.

Aim higher to shoot farther, but keep the relative positions of the rocket, rubber band, and pencil the same.

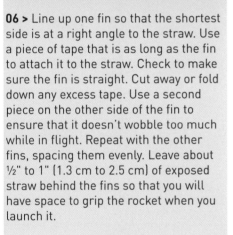

We all know those balsa wood wind-up aircraft kits. Wind up the propeller, place it on the ground and watch it take off . . . and crash! Taped repairs can only go so far before the plane is retired to a forgotten place. But now, the rubber band helicopter can breathe new life into those dilapidated aircraft parts. Salvage the propeller, some balsa wood, and a couple of rubber bands to create a new aircraft that can fly upwards of 20' (6 m)!

RUBBER BAND HELICOPTER

$$\Delta x = \bar{v} t$$

$$\frac{1}{2} a t$$

$$+ v \circ$$

$$_o^2 = 2$$

$$v \frac{v +}{}$$

$$g = 10 \frac{m}{s^2}$$

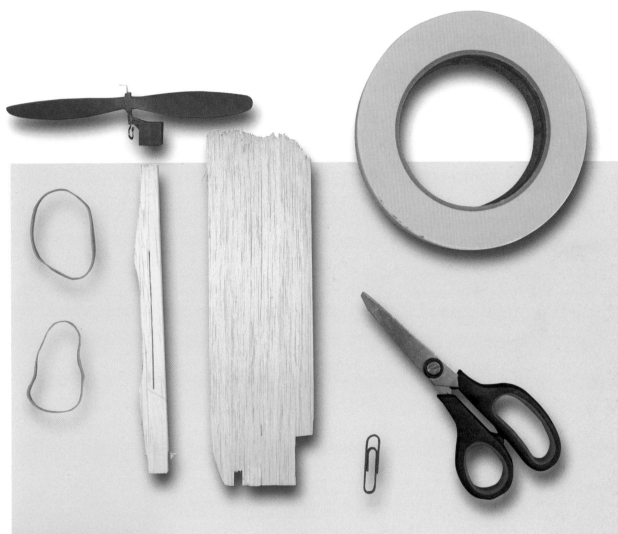

TOOLS + MATERIALS

RUBBER-BAND AIRCRAFT KIT
SCRAP BALSA WOOD (OPTIONAL)
1 PAPER CLIP
SCISSORS
MASKING TAPE
2 RUBBER BANDS

MATERIAL SUBSTITUTIONS

AIRCRAFT KIT	You can purchase plastic propellers from an online vendor listed in Material Sources at the back of the book. Sheets of balsa wood are available at craft stores.
SCRAP BALSA	Lightweight cardboard or card stock

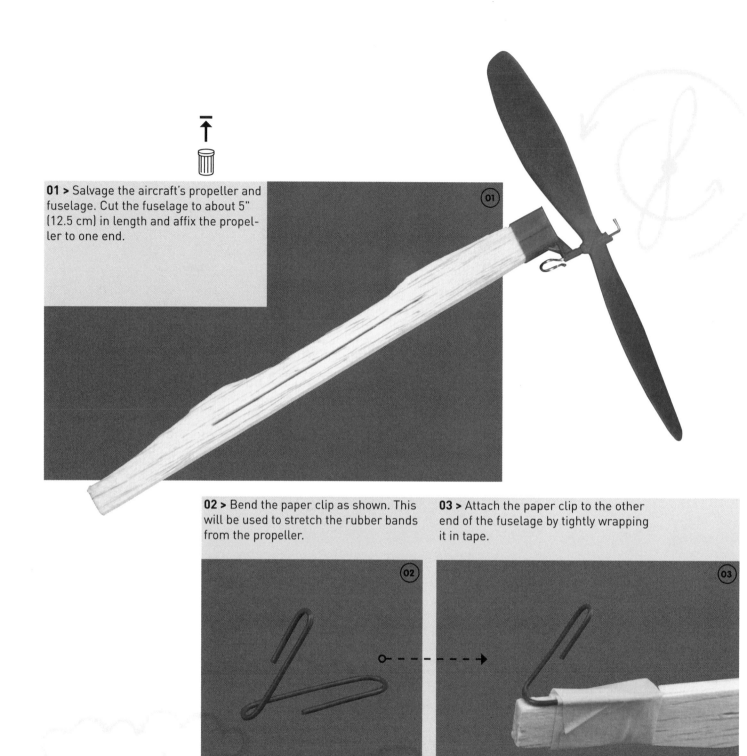

01 > Salvage the aircraft's propeller and fuselage. Cut the fuselage to about 5" (12.5 cm) in length and affix the propeller to one end.

02 > Bend the paper clip as shown. This will be used to stretch the rubber bands from the propeller.

03 > Attach the paper clip to the other end of the fuselage by tightly wrapping it in tape.

04 > Salvage a piece of the aircraft wing and clean up any broken edges with scissors. Cut the piece to about 2" (5 cm) tall and 6" (15 cm) wide. This will provide the crucial component of lateral drag for flight: When the helicopter flies, the rubber bands will spin both the propeller and the fuselage. The propeller provides lift while the fuselage only stretches out the rubber bands. The goal is to have more energy diverted to the propeller. The lateral drag, provided by the piece of wing or balsa, slows the spinning of the fuselage and diverts the energy of the rubber bands to the propeller, allowing the helicopter to fly much higher.

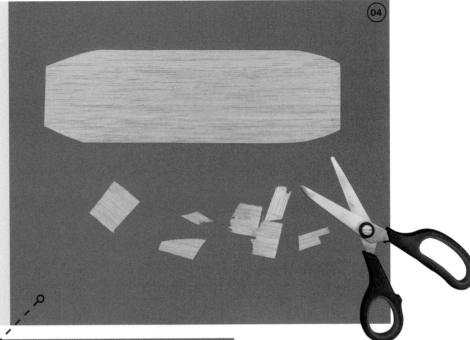

05 > Tape the salvaged wing piece to the narrow edge of the fuselage toward the propeller.

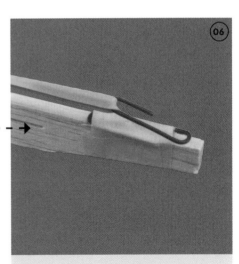

06 > Stretch out two rubber bands to connect the propeller to the paper clip.

07 > Begin winding the helicopter. Use one hand to hold the fuselage, and the index finger of your other hand to wind the propeller counterclockwise.

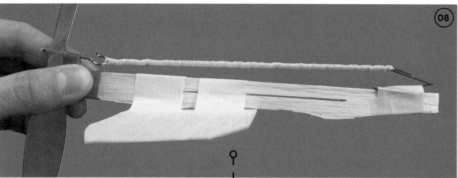

08 > Generating enough energy is crucial for a successful flight. You can gauge how much energy the rubber band has by observing it as it winds. In this picture, the rubber band is completely full of tight coils, but it's not enough to get it off the ground.

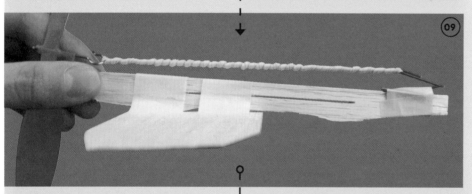

09 > Once the rubber band is lined with coils, it will begin coiling on itself a second time. In this image, most of the rubber band is filled with these double coils.

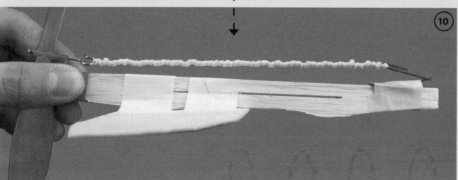

10 > Continue winding until the rubber bands start to coil a third time. At this point, the rubber bands will no longer form a straight line. Now it has enough energy!

11 > To fly: Hold the bottom of the helicopter in one hand and the tip of the propeller in the other. Release the propeller first and then—this is essential—release the bottom of the helicopter within the next second.

If you were to release both the propeller and the bottom at the same time, gravity will immediately pull the helicopter toward the ground before it can start generating lift. Sometimes this results in the helicopter flying in a random direction, and that direction might be toward you!

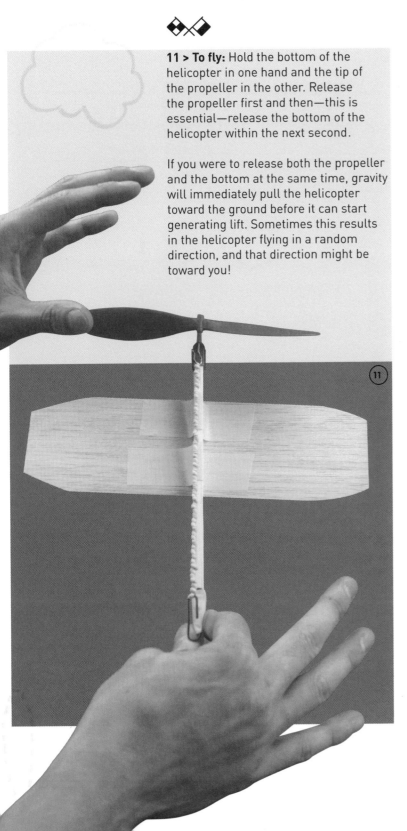

DESIGN VARIABLES

Plastic propellers from online vendors are available in various sizes. See Material Sources in the back of the book for a suggested online vendor. Use one to build a helicopter from scratch with a craft stick for a fuselage and card stock instead of balsa, as shown in the example below.

Cutting the card stock or balsa for the lateral drag of the helicopter is critical to its design and performance. Having a piece that is too big will weigh down the helicopter, and having a piece that is too small won't create enough useful drag. Finding the right balance is key, however there is a wide range of designs that will work just fine.

OTHER GADGETS + CONTRAPTIONS

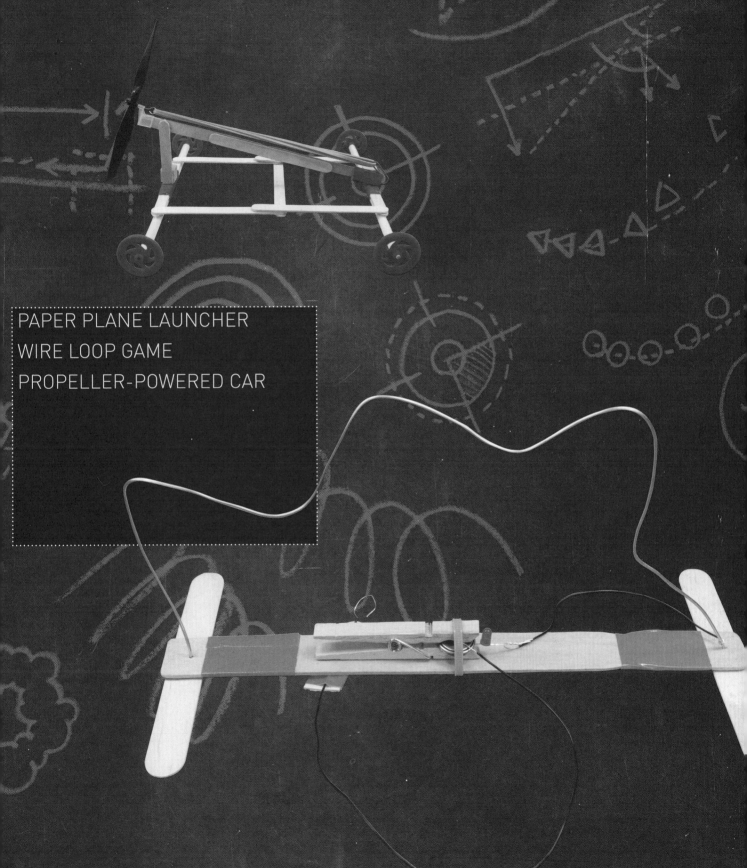

PAPER PLANE LAUNCHER
WIRE LOOP GAME
PROPELLER-POWERED CAR

(i) Transform two handheld battery-powered fans into an automatic paper plane launch pad! By replacing the fan blades with cardboard disks, a paper plane will rapidly launch upward when you feed it between the spinning disks. The launcher can easily fling paper planes a substantial 15' (4.5 m) or more, and you don't have to wear out your throwing arm while using it!

PAPER PLANE
LAUNCHER

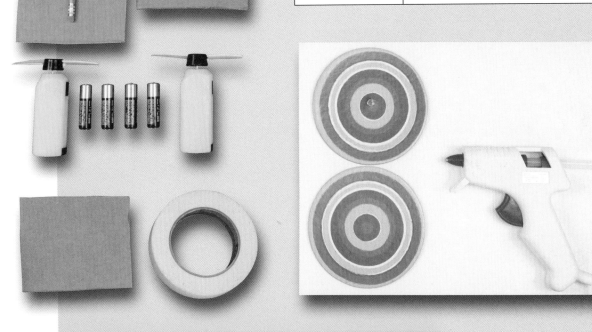

MATERIAL SUBSTITUTIONS	
CIRCULAR CARDBOARD COASTERS	Circular container lids, the bottom of a circular cardboard oatmeal container, small paper plates. Any perfectly round, rigid object between 4" and 5" (10 and 12.5 cm) in diameter. If your material is thin, like a plastic container lid, then you may need to glue two layers together for each disk. Cutting circles by hand is not recommended because very small imperfections will cause the plane launcher to vibrate wildly or self-destruct when the fan motor is at full speed.
DUCT TAPE	**Masking Tape**

TOOLS + MATERIALS

2 BATTERY-POWERED HANDHELD FANS

BATTERIES

2 CIRCULAR CARDBOARD COASTERS, 4" (10 CM) IN DIAMETER

RULER

PENCIL

HOT GLUE GUN

3 PIECES OF HEAVY CARDBOARD

CRAFT KNIFE

DUCT TAPE

SCISSORS

01 > Install the batteries in one fan as you normally would. For the other fan, install the batteries backward. This will reverse the direction of the motor. When you turn the fans on, one should turn clockwise, and the other should turn counterclockwise.

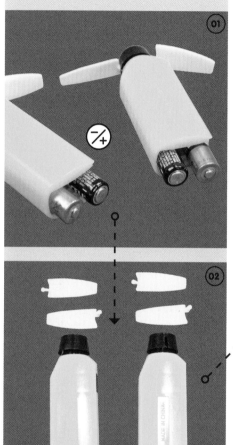

02 > Remove the fans' blades if possible. You can skip this step if the blades are fixed onto the motor shaft or if there is no other area for attaching the circular coasters.

03 > To locate the exact center of each coaster, use a ruler to draw three lines across the edge of the circle, each exactly 3" (7.5 cm) long. Mark the center of each line at 1½" (4 cm). The exact position of the lines is not important.

Using the corner of the ruler as a right angle, draw a perpendicular line from the center of each 3" (7.5 cm) line into the middle of the circle. The center of the circle is where the three lines intersect.

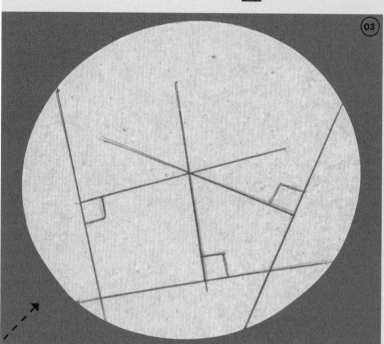

04 > Position the coasters exactly in the center of each motor's axis of rotation and use hot glue to attach them in place. Turn on the fans to make sure the coasters are centered. The goal is to have little to no vibration when they rotate. You may need to remove and replace the coasters if the center of rotation is off by more than ⅛" (3 mm).

The closer the motor's center of rotation is to the center of the coaster, the more effective the launcher will be.

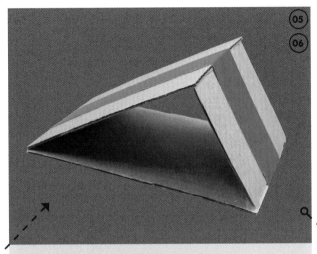

07 > Check the direction of the motors and position the handheld fans on the short side of the launcher. In this picture, the left disk turns clockwise and the right one turns counterclockwise. Hot glue the handheld fans in place. The spinning disks should have about ⅛" (3 mm) of space between them, and should have about ½" to ¾" (1.3 to 2 cm) of clearance above the launch pad. Avoid attaching the motors in a way that prevents access to the on-off switch or the battery compartment.

05 > Create a wedge-shaped launch pad using three pieces of cardboard that measure about 7" × 5" (18 × 12.5 cm), 9" × 5" (23 × 12.5 cm), and 5" × 5" (12.5 × 12.5 cm). Keep in mind that the shortest piece may need to be wider if you are using disks that have a diameter greater than 4" (10 cm).

06 > Hot glue the cardboard pieces into a triangle with the longest piece on the bottom, as shown. It's important that the apex (top angle) of the launcher be a right angle. Wrap the launch pad with a long piece of duct tape.

The glue creates a rigid structure, and the tape will prevent the vibrations from the spinning coasters from breaking the glue bonds.

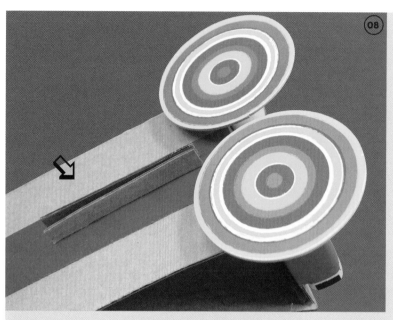

Optional: Wrap a piece of duct tape around both fans. This allows you to tighten the gap between the spinning disks. Loosen the tape if the disks are too close together.

08 > Create the plane guides by cutting two ½" × 5" (1.3 × 12.5 cm) strips of cardboard. Hot glue them to the center of the launch pad, directly in line with the gap between the discs, with about ¼" (6 mm) of space in between them.

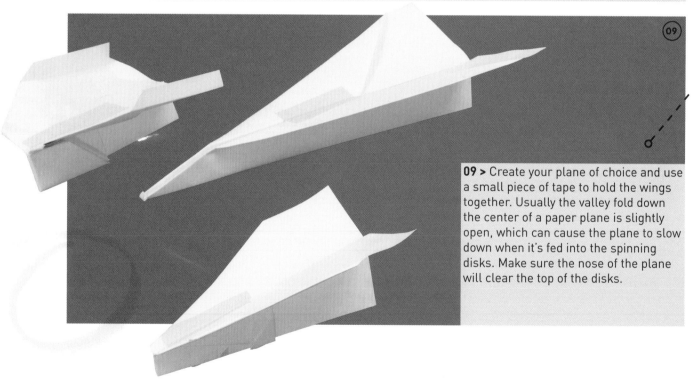

09 > Create your plane of choice and use a small piece of tape to hold the wings together. Usually the valley fold down the center of a paper plane is slightly open, which can cause the plane to slow down when it's fed into the spinning disks. Make sure the nose of the plane will clear the top of the disks.

10 > Prepare to launch! Turn on both motors, wait until they reach maximum speed, and then set your plane between the guide strips. Although handheld fan motors do not have much torque, the accumulated energy in the spinning disks should have enough force to pull the plane through and launch it into the air.

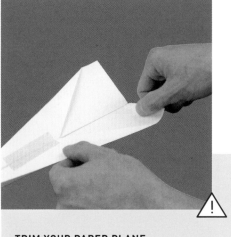

TRIM YOUR PAPER PLANE

Is your plane not cooperating? Try trimming the wings. This means curling the back edge of the wings slightly to correct the amount of lift. Curling the paper upward will increase the amount of lift, while a downward trimming will cause the plane to take a nosedive.

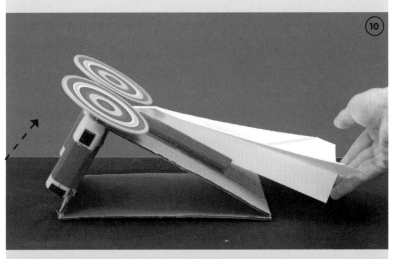

11 > Lightly tap the back of the plane to give it a push into the spinning disks. The disks should grab the plane and fling it up to 15' (4.5 m) into the air!

If you don't find success right away, try adjusting the gap between the spinning disks to match the width of the center of the plane.

WIRE LOOP
GAME

$$\Delta x = \bar{v}t$$

$$\Delta x = \frac{1}{2}at + v_0$$

$$v = at + v_0$$

$$v^2 - v_0^2 = 2a\Delta$$

$$v \frac{v + v_0}{2}$$

ⓘ Precision and dexterity is what this game is all about. The challenge is to maneuver a wand around a wiggly piece of wire without touching it. It's hard to cheat in this game because a simple circuit will activate a bright LED light if the wand touches the wire! The best part about this project is the ease with which you can adjust the level of challenge simply by reshaping the wire.

TOOLS + MATERIALS

1 PAINT STIRRER (OPTIONAL)

3 LARGE CRAFT STICKS (OPTIONAL)

HOT GLUE GUN

DRILL WITH BIT AT LEAST AS WIDE AS THE WIRE USED (OPTIONAL)

THICK AND BENDABLE WIRE (16 GAUGE MECHANICS' WIRE WORKS WELL)

DUCT TAPE (OPTIONAL)

WIRE STRIPPER

INSULATED WIRE (BETWEEN 20 AND 22 GAUGE WORKS WELL)

1 SPRING-TYPE CLOTHESPIN

1 LED LIGHT

1 3 VOLT COIN CELL BATTERY

1 CABLE TIE

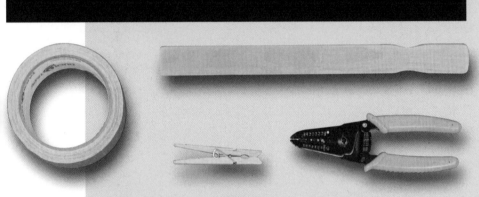

MATERIAL SUBSTITUTIONS	
PAINT STIRRER AND CRAFT STICKS	It's not necessary to use wood for the base. You could build a wire loop game on a shoebox, a piece of cardboard, or any other nonconductive surface.
WIRE STRIPPER	You can strip wire with a pair of scissors and a little dexterity.
LED and 3 VOLT BATTERY	You can disassemble a dollar-store flashlight and use the components from that, however, your circuit might need to be built differently depending on how the flashlight is made.

01 > Build the base for your wire loop game by gluing two craft sticks perpendicularly onto the ends of a paint stirrer to form a wide H shape.

02 > Drill a hole at the center of each end of the base. The holes should be large enough for the thick wire to fit through.

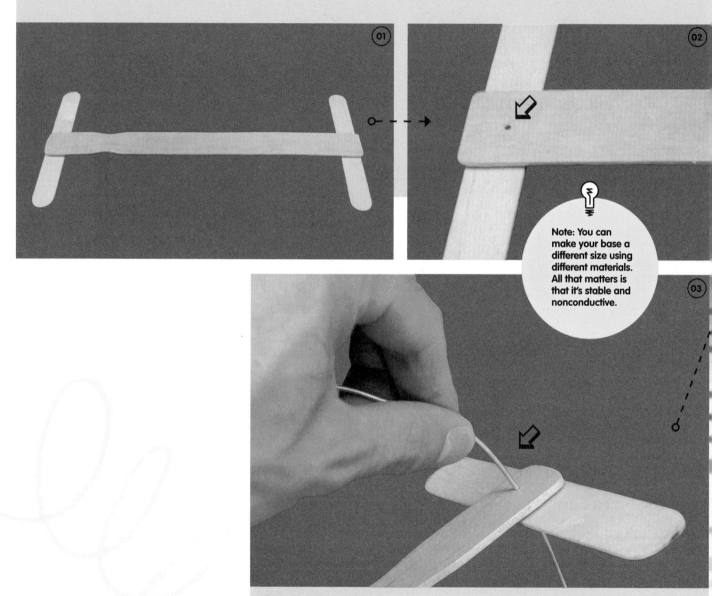

Note: You can make your base a different size using different materials. All that matters is that it's stable and nonconductive.

03 > Insert the wire through the holes.

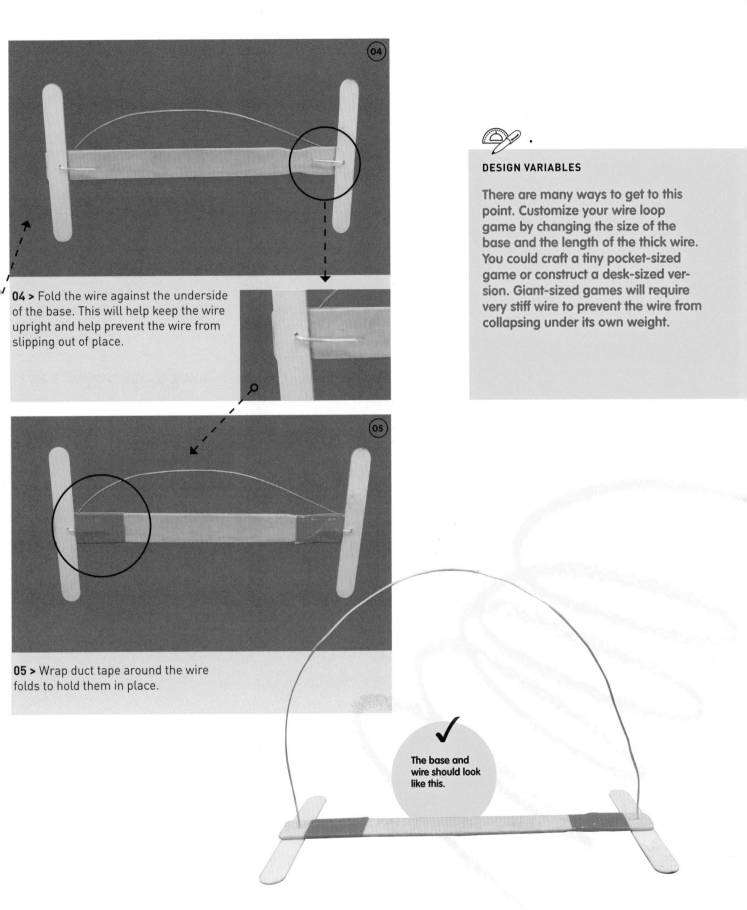

04 > Fold the wire against the underside of the base. This will help keep the wire upright and help prevent the wire from slipping out of place.

05 > Wrap duct tape around the wire folds to hold them in place.

04

05

DESIGN VARIABLES

There are many ways to get to this point. Customize your wire loop game by changing the size of the base and the length of the thick wire. You could craft a tiny pocket-sized game or construct a desk-sized version. Giant-sized games will require very stiff wire to prevent the wire from collapsing under its own weight.

✓ The base and wire should look like this.

06 > Cut a piece of thin insulated wire to about 8" (20.5 cm).

07 > Use a wire stripper to remove about 1½" (4 cm) of insulation at each end of the wire. Alternatively, you can use a pair of scissors to remove the insulation.

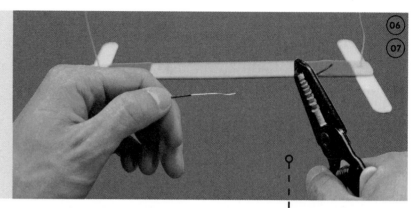

08 > Glue a clothespin to the center of the base. Because this is a solderless circuit, the pinching action of the clothespin is used to clamp the wires to the battery.

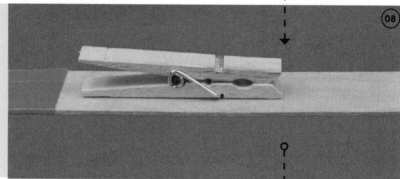

09 > Tightly wrap one end of the insulated wire around the base of the thick wire on each side of the base.

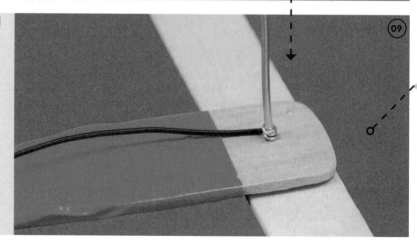

10 > Identify the positive and negative leads on the LED: The positive lead is the longer one.

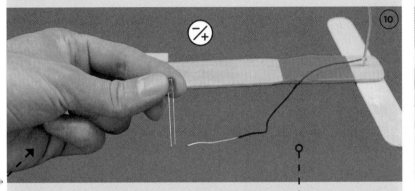

11 > Connect the negative lead from the LED to the insulated wire.

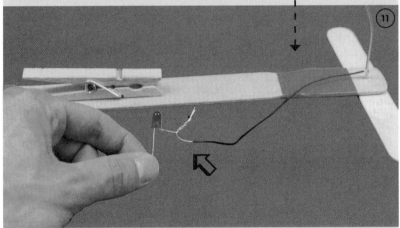

12 > Clamp the positive lead of the LED to the positive side of the coin cell battery. The positive side of the battery has a + symbol and is smooth.

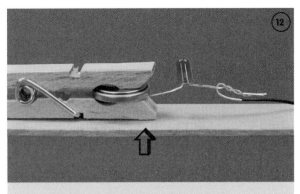

13 > Strip the ends of a longer piece of insulated wire and clamp one end to the negative side of the battery. The length of this wire will depend on the size of your wire loop game. Larger games will need a longer wire.

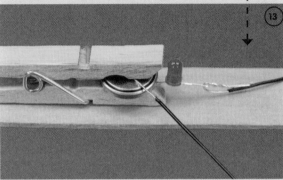

14 > Secure the battery connections by tightly wrapping the clothespin with a cable tie.

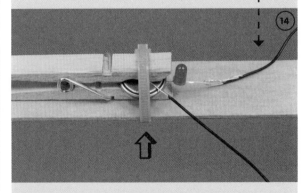

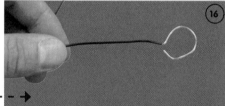

16 > At one end of the long insulated wire, form a small loop that will fit over the thick wire.

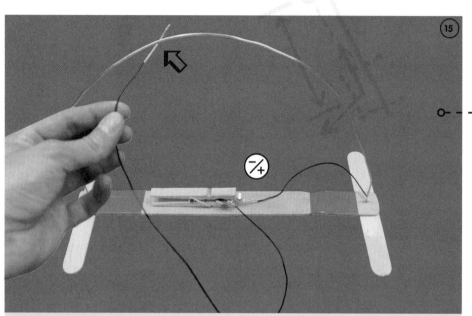

15 > Test the circuit by touching the longer piece of insulated wire to the thick wire. The LED should light up! If it doesn't, try flipping the battery. If it still doesn't work, examine your connections and make sure that the wires near the battery are not touching each other, and that all of your connections have solid contacts.

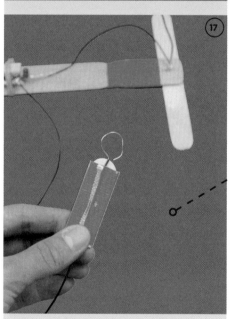

17 > Tape the wire to a half piece of a large craft stick. Be sure to press the tape down firmly to prevent the wire from twisting or moving around. This gives the player something to grip and allows for easy rotation of the loop.

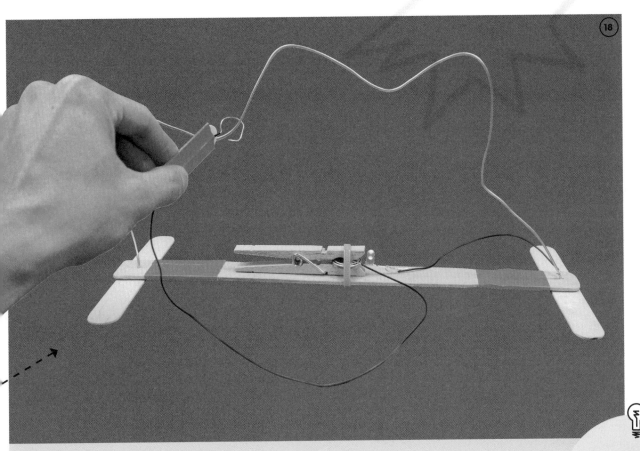

18 > Bend the heavy wire into a more challenging shape, slip the loop over the heavy wire, and try playing! This game is set up in a simple wavy pattern, but you could make it as complex or as easy as you want. Best of all, it's easy to redesign your game for a new challenge every time you play!

When you're not using the game, tuck the loop away to prevent it from accidentally activating the circuit and burning out the battery.

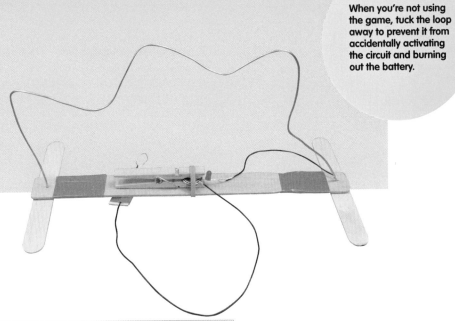

Tired of flying your rubber-band helicopter? Maybe you have two broken plane kits? Repurpose those parts into a speeding desktop vehicle! All you need is a pair of improvised wheels and 30' (9 m) of runway.

PROPELLER-POWERED CAR

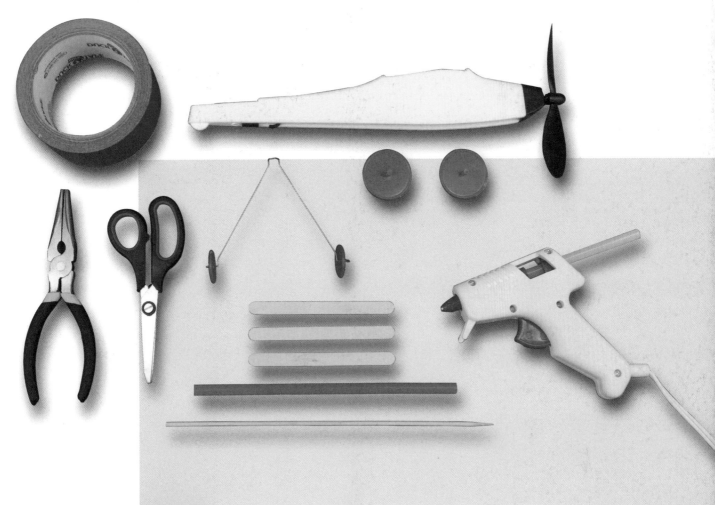

MATERIAL SUBSTITUTIONS	
PLASTIC TOPS	These will be the extra pair of wheels. You can make alternative wheels by drilling holes into plastic bottle lids or tracing a circle onto cardboard and cutting it out.
	Regardless of what you choose, make sure that the hole is centered and that it's slightly smaller than the diameter of the skewer.
CRAFT STICKS	Anything that can hold the back of the car off of the ground.

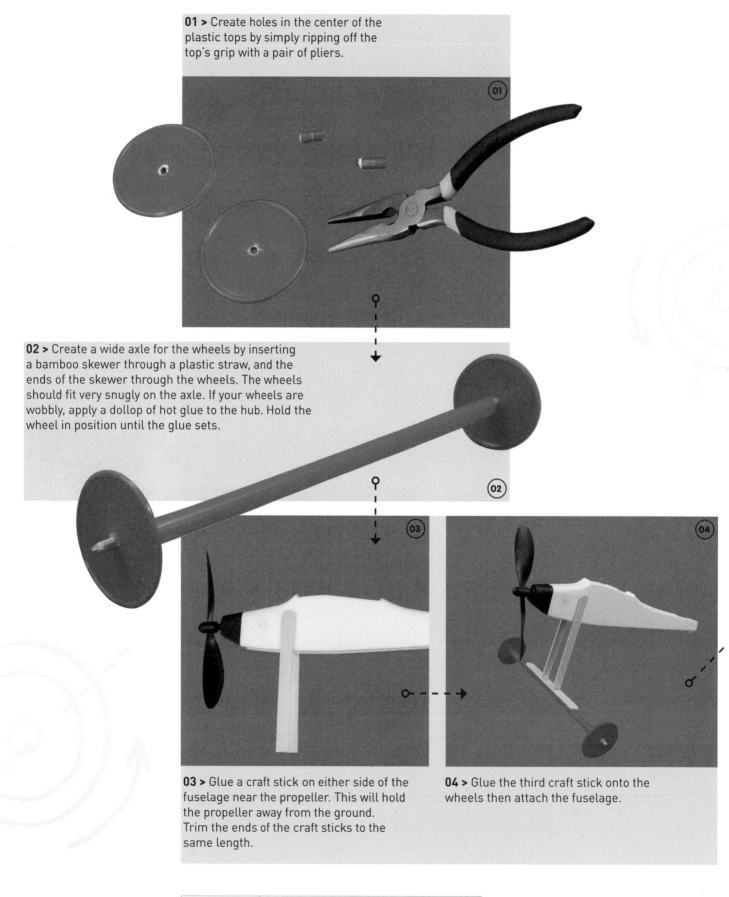

01 > Create holes in the center of the plastic tops by simply ripping off the top's grip with a pair of pliers.

02 > Create a wide axle for the wheels by inserting a bamboo skewer through a plastic straw, and the ends of the skewer through the wheels. The wheels should fit very snugly on the axle. If your wheels are wobbly, apply a dollop of hot glue to the hub. Hold the wheel in position until the glue sets.

03 > Glue a craft stick on either side of the fuselage near the propeller. This will hold the propeller away from the ground. Trim the ends of the craft sticks to the same length.

04 > Glue the third craft stick onto the wheels then attach the fuselage.

05 > Salvage the landing gear wheels from the plane kit and tape them to the front of the car. I used a piece of duct tape cut in half lengthwise.

06 > Widen the salvaged landing gear for added stability. The propeller-powered racer is finished! Power it with the rubber bands that came with the kit, or experiment with your own ingenious elastic combination.

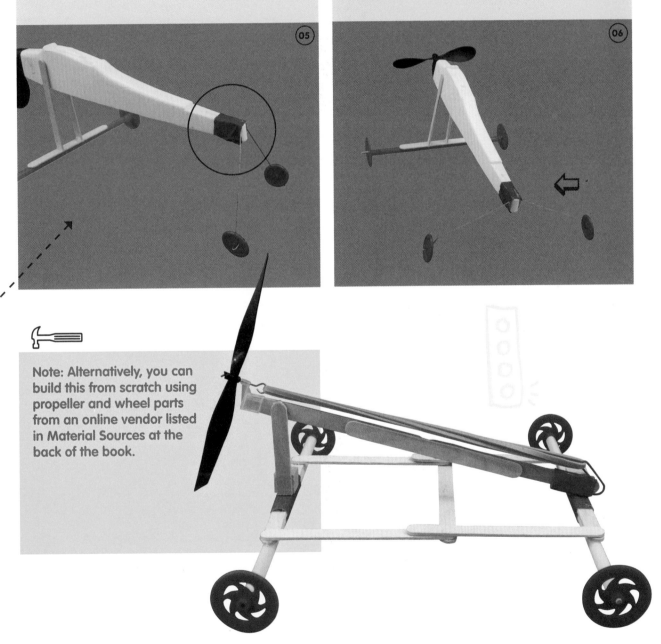

Note: Alternatively, you can build this from scratch using propeller and wheel parts from an online vendor listed in Material Sources at the back of the book.

MATERIAL SOURCES

The materials used in this book are inexpensive and easy to come by. Materials not listed here can most likely be found at hardware and department stores or in your kitchen junk drawer. Note: Online vendors may not ship to all countries.

CRAFT CUBES
Cubes with holes and dowels of many varieties can be purchased at www.CraftParts.com.

CRAFT STICKS
Find crafting sticks at craft or dollar stores.

HOT GLUE
Available at crafting, hardware, or department stores. I recommend buying the glue sticks online or at dollar stores to save money.

LEDS AND 3 VOLT COIN CELL BATTERIES
Purchase these at electronics supply stores or through online vendors.

LONG RUBBER BANDS
These are available by the box through office supply stores.

PAINT STIRRERS
Paint stirrers are available at hardware stores, sometimes for free if you ask nicely. You can also purchase them online. I bought boxes of 100 for this book for about $20 USD.

PLASTIC SYRINGES
Pharmacists may give a few to you for free if you're a regular customer, or if you ask for some for a "science project." Otherwise you can buy them online. Search for "10 ml luer slip plastic syringe."

PROPELLERS
Propellers for rubber-band helicopters and propeller-powered cars can be purchased at www.Kelvin.com. Search for "nose hook propeller."

VINYL TUBING WITH ⅛" (3 MM) INNER DIAMETER
This tubing is available at aquarium supply stores, hardware stores, or online.

ABOUT THE AUTHOR

Lance Akiyama combines tinkering and education into one aspiration: to create a better world by inspiring the next generation of innovators with exciting hands-on projects.

To that end, he has created many project-based learning tutorials on Instructables.com, started a small, after-school engineering service, and is currently employed as a STEM-based curriculum developer for Galileo Learning.

Other than that, Lance spends his free time designing elaborate plans for advanced contraptions, keeping journals in cryptic backwards writing, and attempting to fly by strapping paper wings to his arms and leaping from rooftops.

INDEX